D0427771

Praise for

The Geese of Beaver Bog

"Heinrich's lyric writing and attentive observations . . . make the goose world come alive. . . . Reading *The Geese of Beaver Bog* . . . is pure joy and the next best thing to taking a deliciously long and tranquil bog outing of one's own. . . . It is a respite, a delightful sanctuary in this world that seems to grow more chaotic and noisy by the day."

—*Los Angeles Times*

"Fascinating. . . . A saga full of drama. . . . Heinrich is an extraordinary writer, and an impressively disciplined and rational observer. The power of the book comes, however, from the emotion that he very deftly allows the reader to recognize and share. . . . A soaring narrative full of charm and affection as well as precision of reporting."

—*Baltimore Sun*

"Heinrich aims to recreate the rhythms of bog life. . . . The intent, as in Heinrich's earlier books, is to apprehend an animal from the 'inside,' and so learn implicitly how it knows what it knows, why it does what it does. If geese were literate, they'd no doubt offer encouragement after reading Heinrich's account. 'Yes,' Peep might honk, 'that's how it was.' "

—*Natural History*

"Arguably today's finest naturalist author. . . . This is another worthy missive from our latter-day Thoreau."

—*Publishers Weekly*

"[Heinrich is] an astute observer of nature. . . . There is much that students of animal behavior will learn by reading this insightful naturalist's thoughts on the personal lives of geese."

—*American Scientist*

"*The Geese of Beaver Bog* is as much a loving memoir as a naturalist's log. . . . [Heinrich's] writing, with its detail-oriented first person narrative, evokes a gentler time. . . . Charming, both personal and warm."

—*San Francisco Chronicle*

"By the end of the book, these highly individual honkers seem like our own family."

—*Boston Globe*

"A nice introduction to animal behavior. . . . Heinrich writes beautifully . . . [and] his powers of observation are impressive."

—*New Scientist*

"Underlying the engaging, personal nature of the narrative is Heinrich's scientific background, and the reader learns quite a bit about marsh biology and goose behavior between the lines."

—*Booklist*

© WILLIAM LIPKE

About the Author

BERND HEINRICH is the author of numerous books, including the bestselling *Winter World*, the award-winning *Mind of the Raven*, *Why We Run*, *The Trees in My Forest*, *A Year in the Maine Woods*, *Ravens in Winter*, *One Man's Owl*, and *Bumblebee Economics*, which was nominated for the National Book Award. A professor of biology at the University of Vermont, Heinrich divides his time between Vermont and the forests of western Maine.

THE GEESE OF BEAVER BOG

Also by Bernd Heinrich

The Trees in My Forest

A Year in the Maine Woods

Ravens in Winter

One Man's Owl

Bumblebee Economics

Mind of the Raven

Why We Run
(previously titled *Racing the Antelope*)

Winter World

ecco *An Imprint of* HarperCollins*Publishers*

Bernd Heinrich

THE GEESE OF BEAVER BOG

 For information, address HarperCollins Publishers Inc., 10 East 53rd Street, New York, NY 10022.

HarperCollins books may be purchased for educational, business, or sales promotional use. For information, please write: Special Markets Department, HarperCollins Publishers Inc., 10 East 53rd Street, New York, NY 10022.

All line drawings and photographs are by BERND HEINRICH.
The map on page 103 is by LAURA HARTMAN MAESTRO.

First Ecco paperback edition published 2005

Book design by Shubhani Sarkar

Library of Congress Cataloging-in-Publication Data is available upon request.

ISBN 0-06-019745-5
ISBN 0-06-095738-7 (pbk.)

05 06 07 08 09 ❖/RRD 10 9 8 7 6 5 4 3 2 1

Acknowledgments

To Rich Howard and Kent Wells for information on frog chorusing; Paul Sherman for discussions and reprints on nest parasitism; William Crenshaw for information on Canada geese; Liz Thompson for sedge identifications; Louise O'Hare for typing; Dan Halpern, Gheña Glijansky, and Diane Aronson for editorial comments and suggestions; and Theora Ward, our neighbor, for allowing me free access to the best beaver bog within a stone's throw of our driveway.

Contents

Introduction

There is something in the ceaseless chatter of migrating geese that stirs me. Perhaps it touches something wild, remote, and mysterious that I share with them, for it is almost with longing that I look up every fall and spring when the scraggly V formations wing their way overhead high in the sky. Perhaps it is the tenor of their haunting cries, their mastery of sky and distance, their commitment and single-mindedness in striving to reach far-off goals that enchant. Whatever it might be, the migrating geese are a phenomenon of nature, even though to the naked eye they are barely visible patterns of dark specks against the endless skies.

For many years the geese seemed physically remote to me. That changed dramatically after my family and I got to know an individual Canada goose. We called her Peep. Peep was not a phenomenon, and I can't say she inspired. She became too close and personal to us for that. Yet, she always retained a connection to the wild, and I hope she still is living the life that wild geese live.

Peep had stayed with us for a while one summer a few years ago, and through her I was introduced to other Canada geese who nested in a nearby beaver bog. Her inadvertent gift to me was that through her I came to know other geese and other swamp denizens, especially the common grackles and the red-winged blackbirds who were a study in contrasts.

Peep was a large bird. Her long and gracefully curved neck was jet-black and felt velvety to the touch. Like other "honkers," as Canada geese are affectionately called, her face was adorned on each side with an immaculate white patch that had, if you looked closely, a unique individually identifying pattern. Her prominent dark brown eyes, located just above the upper border of the white face patch, seemed ever alert. The gray-brown feathers of her back were each fringed with light tan, giving them a scalloped appearance. She kept her great wings neatly folded and tight against her back. They shielded her from the cold, the rain, and the sun, which was most of the time. Like all geese, she was equally adept in the sky (except during the summer molt when she could not fly), on the water, and on dry land. Her legs were set widely apart, which gave her a gentle swagger when she walked. When on the water, she paddled her feet alternately and drifted forward smoothly. A tiny kick of her foot underwater produced much forward thrust, because while a goose's toes compressed laterally on the forward swing of the foot, they expand laterally to expose the webbing between the toes on the backward thrust.

From the time that she was a yellow fluffball hatched out of a big, beige egg until she grew into a splendid specimen of adulthood, the one word that described her character (in human terms) was "sweetness." When she was still young, that adjective referred mainly to her looks and her endearing peeps. As she grew older, it referred to her personality, which contrasted with the bold and aggressive nature of many a gander. She grew up to be reticent, but she never showed a sign of fear.

Peep's face.

Like all geese, and unlike many other birds who tend to be nervous, jumpy, and continually excitable and excited, Peep usually maintained a steadfast calm. She was alert, but moved in a measured deliberate manner. She could show excitement, but was slow to excite. She could sound the Klaxon, but she usually stayed silent or communicated with me in barely audible chuckles and grunts.

Our family came to know Peep as a gosling, and in effect we became parents to her. After she came of age and left us, I performed a daily ritual of accompanying her and other geese in the beaver swamps near the foot of our driveway, at our home close to Burlington, Vermont. I tried to enter her world of potential mates, rivals, and other animal neighbors who shared the bog. I watched fights, the ritual of choosing a nest site, mating, the hatching of eggs, the drama of the young leaving the nest, and in general, the entire process of their growing up. I saw parenting styles and personalities, and contrasted both with those of other birds living in the same environment. I gave some of the geese that I knew human names, not only in order to make it easier to remember them as individuals, but also because they reacted to each other as individuals themselves.

This is the story of their lives as I observed them through Peep's life and through a biologist's lens. It is important to note, however, that the focus of this book is not on "hard science." Instead, its focus is on individual geese—each possessing a unique history and a particular set of relationships with other individual geese. It would be difficult to tweak my very specific, often singular, observations into the framework of "hard science." And I would not wish to do so.

There has been progressive improvement in our collective understanding of animals, but the methods of science have hindered as well as promoted this understanding. For most of the last two centuries, scientists considered animals other than ourselves to be little more than reflex autonoma, simple creatures that could be conveniently studied in boxes. In fact, what made an animal a candidate for study was its suitability for being observed under tightly controlled lab conditions. "True" science was defined as that which can be replicated ad infinitum by any experimenter.

Those who studied behavior in the field were until recently considered to be doing "soft" science. It took some time before it was officially acknowledged that an animal is not really whole until it is within its natural environment, and that observations made under laboratory conditions necessarily miss crucial factors even as they illuminate others. This, and the realization that behavior evolves, spawned a discipline called *ethology*. Karl von Frisch (bee communication), Nikolaas Tinbergen (instinct in seagulls), and Konrad Lorenz (imprinting and greylag geese) shared a Nobel Prize for pioneering this new approach. Ethologists sought to discover species-specific, genetically encoded "rules" (i.e., common denominators) that would—as all rules must—apply uniformly. In that way, they were not altogether different from the "hard" scientists. The problem is that rules are, by definition, simple—and animals are not. Animals are also genetically unique as individuals, and behaviorly variable.

So-called biological rules are the sum of individual cases, and an observer of individual cases can be treated to surprising, anomalous observations. Some of them may become a beacon to discovery. That was the case for me with Peep and her associates.

THE GEESE OF BEAVER BOG

1. IT'S A GOOSE

The speed limit on the highway a mile from my home in Vermont is 45 miles an hour, and Peep was pushing it. She was winging along a foot or two behind and just to the left of the cab of my Toyota pickup truck. Another truck roared by from the other direction but she kept her place. She didn't miss a wing-beat. You might think she knew all about flying, road vehicles, and the right-of-way convention when barreling down the highway. Fact is, this was her maiden flight. I was as surprised to find her beside me as I suspect the truck driver was.

I had originally planned to drive to town, but seeing her now I reconsidered. I slowed, turned around, and headed back toward our dirt road to lead her home. She was soon again with me. I then cranked up to 50 miles an hour on the last level stretch on the approach to the turnoff to our road, to see what she could do. She started to lag a bit and I knew she was pushing, approaching her limits, because her bill opened and as I glanced sideways I saw her pink tongue exposed while she panted from exertion and overheating.

She didn't turn the corner too well. Tongue still out and chest heaving, she landed in a ditch and waddled out onto the dirt road. I stopped to see if she was all right. After giving her a couple of minutes to catch her breath, I got back into the truck. As I drove off and looked into the rearview mirror I saw her running behind me, then flapping her wings and again becoming airborne. And so we came back home and went up the long driveway through the woods to our house opposite our beaver swamp.

I had known Peep since a couple of days after she hatched from a big cream-colored egg. She had grown up and lived all that summer of 1998 on our lawn. Except for a sibling who lived for only a few days after she had hatched, Peep had never known other geese. But geese are social animals. She followed me and other family members like she might have her own, which made it difficult for us to go anywhere without having her in tow. On this afternoon in late September I had tried to trick her, in order to make my escape, by dropping a handful of cracked corn at the back door and dashing into the house and then coming out the front door to quickly jump into the truck. This time, unlike most others, she had left the bait almost immediately to go look for me, and she had seen me enter the truck.

To my surprise, she got excited when I started the motor. She ran back and forth in front of, beside, and under the truck, rather than being frightened off as I had expected. Apparently the clamor of the motor attracted her. I had entered something big and noisy. According to the ethologists Konrad Lorenz and Nikolaas Tinbergen, many animals respond with "fixed action patterns" (FAPs) to specific "sign stimuli" that represent relevant features of their world. Perhaps to Peep I had entered something akin to a flock; I had disappeared into something big, mobile, and noisy. And so she had followed. It was near the time that geese start their northward migrations or follow their elders, who lead them to traditional feeding and overwintering areas.

About a week after our adventure, I, my wife Rachel, and our small son Eliot, made a rare getaway over the whole weekend. When we came back, our Peep was gone. We missed her. We had become more fond of her than we thought possible. However, her disappearance was expected, even feared.

My interest in Canada geese had, until then, been peripheral to swamp-watching. I had occasionally seen or heard wild geese visiting our beaver bog since we moved here in the mid-1980s. Vagrant

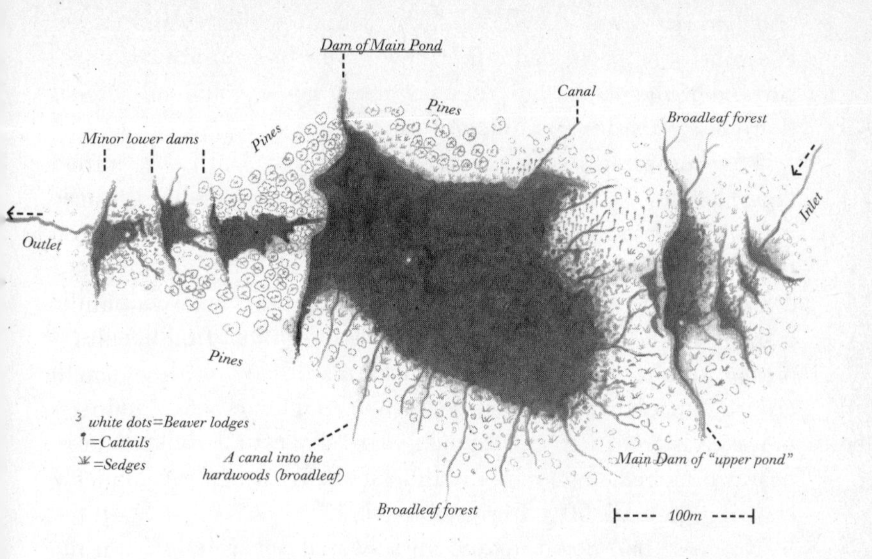

Aerial view of my beaver bog system showing the succession of beaver dams and the main beaver lodges. Only the main (lower) and upper ponds were used by the geese. (Sketches of the beaver dams holding these ponds follow.)

geese may have come by and landed on the pond occasionally, but I did not notice. Then, one dawn in early April 1997 I did notice: I heard geese honking down by the pond. They woke me out of a sound sleep.

The geese's calls in our swamp excited me because they were loud and agitated, as if something was up. Like a goose responding to a powerful sign stimulus, I jumped out of bed, rushed down the driveway, and crashed through the bushes to the edge of the pond, where the ice had just melted. Far out on an open patch of water I could make out a *pair* of geese chasing a third one, and I immediately wondered if the pair might be acting aggressively because they intended to nest here.

Throughout the rest of the month I saw the pair often in the day and heard them at night. My hopes soared: they seemed determined to stay. Then, on the afternoon of April 29, 1997, I saw only *one* goose, and this goose acted differently than any I had seen before. As I walked down through the woods to the pond, it saw me and put its head down to slink behind some dead cattails. Previously the geese had either flown away or swum to the opposite end of the pond when they saw me. A goose would only *hide* if there was something *to* hide. A nest?

I was probably seeing the gander, I thought. Until now he had kept close by his mate's side and so it seemed possible that the goose (female) might at this very moment be somewhere sitting on eggs. I scanned the bog but could see no sign of her, and so I climbed into the top of a large sugar maple tree in the woods at the edge of the bog to get a better view. I still couldn't find her. I saw only a number of red-winged blackbirds displaying their bright crimson epaulets and challenging their neighbors with loud *oog-la-eee* calls from the tops of last year's dry cattail stalks. A lone beaver hunched down in the middle of the redwings' cattail patch, chewing the bark off a fresh willow stick.

No geese had flown up and geese would not walk off into the surrounding forest, so the goose had to be somewhere below me. I continued to look, entertained by the antics of the redwings and enjoying the view from my perch on a thick limb. I saw several old beaver dams at both sides of the drainage between the hills. Dark water from the large main pond spread out below me, and a peninsula of sedge hummocks with the past year's beige-tan, dead, matted sedge foliage jutted out into this water. I saw the old beaver canals that had been created over the years where trees and brush had been dragged from the woods and brought in to the pond. And then I saw a movement of a patch of white among sedge hummocks along the jagged contours of the eastern shore of the pond. It was the white face patch of a goose whose belly was pressed down even as she was pulling vegetation toward and under her. I had found the nest!

This bog, swamp, or marsh (it has features of all three monikers) would now take on a new meaning. It would connect me to those unreachable great birds that every fall and spring fly and cry so hauntingly in their great and mysterious formations high in the sky. My interest in the swamp ratcheted up a notch or two because its liveliness had increased as a result of the pair of honkers who had chosen this place to settle in and rear their young.

As I continued to watch, the gander eventually reemerged onto the pond from behind the cattails. But when I started my descent from the still-bare maple tree, he saw me and hid anew even before I forced my way through the *Viburnum* bushes alongshore. Then I waded into the ice-cold water among the sedge hummocks, and the goose draped herself flat over the nest. When I got to within about twenty feet of her and finally caught her eye, she jumped off and flopped onto the water and started to call loudly. Although the gander had remained hidden until then, his mate's frenzied alarm had immediately brought him out of hiding. He flew in and splashed down next to her, joining her wild clamor.

The nest was made of last year's dry sedge leaves scraped together into a mound. It was surrounded by black water and old beaver canals. Four huge yellowish-green eggs were cradled in a shallow depression in the center of a nest that seemed surprisingly small for such large birds. One of these eggs was immaculately clean, but the other three were vegetation-stained, suggesting that bog water had wicked up into the nest and that the goose had turned the eggs. I suspected that the clean egg had just been laid and I dipped it into the water. As expected, it sank quickly, as fresh eggs do and rotten and live incubated eggs don't. As I gently replaced it into the nest, I took care to avoid abrupt movements and to talk soothingly to the agitated geese. I kept up my soft chatter, even as their wild cries some fifty feet from me continued. Their normal predators (wolves, foxes, skunks, coyotes, and raccoons) would skulk in silence, which I suspected would be behavior tagging them through sign stimuli that spell "predator." I wanted to be anything but surreptitious. I wanted these geese to see that even though I had approached the nest I would not harm it, so they might become as habituated to me as they were to the beavers. The goslings would learn to identify enemies from the reactions of their parents. I wanted these geese to teach their goslings that I'm no threat, so I could eventually see them free on the pond without their immediately scrambling away from me and hiding.

I returned to the nest (for the third time) on May 27, twenty-nine days later. As twice before, the female slipped off the nest as

I got close, and I again talked soothingly to her and to the gander who came to stand by. This time both were at ease with me. Ordinarily when a bird has invested well over a month into its annual reproductive effort (especially when there is no longer time to start over), it is considerably more defensive, not less. Apparently the geese were habituating to me.

When she slipped off the nest this time, I was stunned with surprise and wonder. Instead of four eggs I saw three of the cutest yellow down-puff goslings imaginable. One rotten egg (a floater that stank) remained. I had an urge to scoop up one or more of the babies and take it home with me. But good sense prevailed; I let them be. What could I possibly do with geese? I expected to be able to be near them every day throughout the summer as the family became ever tamer, revealing intimate details of their life.

But it did not turn out as expected. The next day the whole family was gone. The young normally leave the nest in a day or two, but I had expected them to stay on the pond. I looked closely and also the next day, and the next. . . . No doubt about it—the geese had vanished. For the rest of the summer I did not see or hear another goose on the pond.

I shifted my interest to the red-winged blackbirds, to grackles, and to the beavers' behavior and its ecological ramifications in the creation of the bogs with willow thickets, alder swamps, cattail marshes and ponds. With a possible book in mind about beavers and other keystone organisms in northern ecosystems, I examined the new and old beaver works up and down the valleys in between the hills where we lived. The beaver productions with their ponds hosted numerous redwings and I usually found a pair of geese, and a few muskrats who had built their own lodges in the beaver bogs. The geese often choose either a beaver or a muskrat lodge as their nest site. Such dens are islands surrounded by water. They exclude foxes, coyotes, skunks, and raccoons during the birds' most vulnerable stage: the incubation of their eggs. I wondered if one of these predators had taken the goslings that had wandered onto shore. But at a neighbor's pond, also with a tiny "island" that served as a nest site, I commonly saw numerous geese and their goslings at the pond edge. They grew and thrived, even with cows, and occasionally dogs nearby.

Rural Vermonters live close to wildlife and many have livestock. A neighboring farmer kept four domestic geese. Eventually a coyote killed two of his ganders. He shot the coyote, and penned the two remaining female geese. Every spring they laid their eggs into the same nest and incubated them, and every year the female "pair" had to give up their nesting attempt because the eggs were infertile. One year later (1998) the farmer placed two hopefully fertile, wild goose eggs among the geese's own. This time the pair's incubation effort paid off; two goslings hatched and when our then two-year-old son, Eliot, laid eyes on them, he instantly "had" to have them. The farmer was obliging, and in the custody battle between the geese and boy, the latter won out.

Peep. Just hatched (and dried).

My worst fears about becoming a surrogate mother goose were realized. It is trying enough to continually keep track of an active human toddler, but keeping tabs on a toddler taking care of two hatchling geese is more challenging. A gosling's life without its mother or surrogate mother and the protector gander is about as safe as that of a june bug in the chicken coop. Geese are born acting like they know this. The goslings stay continuously within a foot or two of their parents or adopted parents, and if separated by even three feet from them they run to catch up. Failing that, in a second or two they summon up continuous long, drawn-out peeps that to us sound like the heartrending cries of a baby. They are impossible to ignore. As you resume contact, the goslings immediately change their tune. They eagerly run up to you, lower their heads, and show their unabashed pleasure to be with you again by a rapid high-pitched chatter of greeting. Now everyone is happy. Normally, of course, the goslings don't stray far or else every predator for miles around would be alerted. Unless you cater to them instantly, they give you not one second of peace. In short, goslings engage in one of the more effective blackmails on their parents or foster parents that nature can devise.

Humans also respond powerfully to sign stimuli, and the instinct of a human toddler, especially one who is homozygous for "the naturalist gene," is to try to get its clutches on anything (1) cute and fuzzy, (2) moving, and (3) noisy. Baby goslings fit the bill perfectly. Since a boy adores his little pets, he is all the more reluctant to let go once he has managed to get hold of one of them. But love, devotion, and caring do not necessarily translate into the object's well-being, unless fortified by ample knowledge gained through practical experience. The latter often comes slowly, and painfully, to all concerned.

I will leave most of the details of that summer to your imagination, except to point out some highlights. The goslings spent much time next to the glass porch door by the lawn where they liked to graze on grass and clover, and a generous layer of guano soon accumulated at our front door. One of the goslings, appropriately called Poop, quickly succumbed, possibly after eating something he shouldn't have, although we were never sure about that. The other, called Peep, survived the summer with only a cut on her left eyelid that left a small scar.

By fall Peep had become slightly more independent. She was sometimes content to move more than a few feet from a member of the family without protesting in her pitiable cries. However, she would still peep like a little gosling when I was not near enough, and she continued to greet me eagerly. If I needed to get away from the house, I now dumped a pile of cracked corn at the back door, went inside, and then escaped out the front of the house. But as already indicated, she got wise to that. We never saw her fly except for the one aforementioned time that she surprised me after I drove off in the truck.

Her disappearance was not unexpected. I hoped she joined the migrant geese who were then honking loudly and flying high overhead in their formations of hundreds, but I feared worse.

The following summer (1999) after Peep had been gone for a year, I again saw a pair of geese on our beaver pond. The geese were becoming visibly tamer by the year, so I suspected it was the same pair I had first seen in 1997. They nested on a beaver lodge in the main beaver pond as they had the year before, and this time they successfully hatched five young.

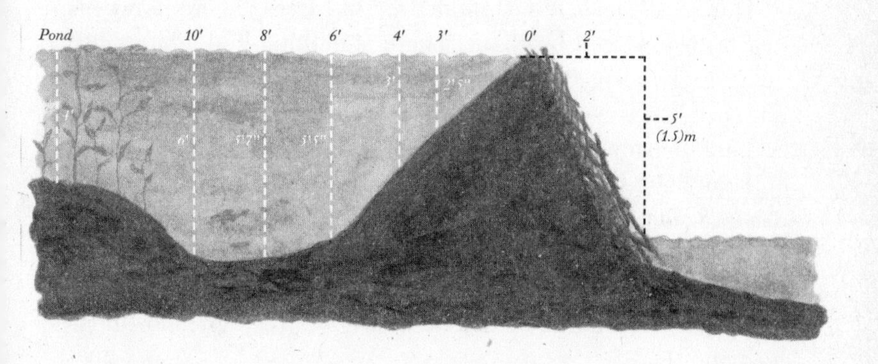

The well-maintained dam of the main pond.

Until the goslings arrived, I saw the geese every day. Strangely, the day after the goslings hatched the family *again* disappeared. I again saw no more geese on the pond just as had happened the year before. I had no idea what happened. I had not disturbed the nest and it seemed unlikely that I had scared them off. The mystery of the disappearing geese deepened.

In early March of 2000, two years after Peep's disappearance, a (the?) pair of geese came back to reside at the pond again. By now I had grown used to them and they to me. They were no longer alarmed by me since I came down to the bog every day at dawn and sometimes also later in the day. I was still interested mainly in the beavers and their work and effects in the area. The geese were there all the time, and we had reached a level of mutual tolerance that bordered on dismissal.

Later that month, on the twenty-eighth, I heard, as usual, the resident geese noisily cavorting down on the pond at night and early in the morning, presumably as they chased one of their many rivals out of their pond. But on this early dawn I awoke and sat bolt upright in bed, because I heard geese honking and *these* geese were flying around our house. Although the pair who came back every year to the bog were by now tame, they were not so tame that they would visit. They had no reason to go out of their way to come all the way up to the house.

Our small lawn is surrounded by tall trees. There is no water nearby. Not even a birdbath. There is nothing here of interest to geese. . . . Unless—?! I jumped up and went to the bedroom window, just as a noisy goose was making another pass. Then the excited honking stopped. Total silence.

I ran downstairs and looked out, and there, right on the gravel driveway by the front door, stood a pair of Canada geese. Our driveway is hemmed in not only by woods but also by the house. Geese don't like such enclosed spaces. What was this pair doing *here*? Why did they land? There was not yet a sprig of green grass for them to feed on.

One of the pair, the larger and from his stance and attitude the gander, was standing tall with his long neck stretched straight up. He was quite still and looked tense. The goose was smaller. She in contrast seemed relaxed and animated. She waddled about with her head and neck in a graceful arch. She stopped and glanced toward the porch several times.

From the instant I had heard them flying around the house, I had imagined that it might be Peep returning. However, I caught myself—this was wishful thinking. She had been gone for so long—two winters—that I could not indulge in such fantasies. Even if it was her, Canada geese normally don't nest until their third year. Furthermore, Peep had been imprinted on humans by us. Everyone knows such animals are not fit for the wild. Would she ever know how to join up with a *goose*? Or would another goose,

Pop on watch while Peep sleeps.
Stance and attitude—him vs. her.

one who did not know what a proper mate looks and acts like, join up with her?

But as I watched this goose who was so fully at ease, I suddenly *knew*. It *was* Peep! And she had a mate with her. To see her now, after two years of thinking she must be dead, was an unbelievable rush. I did not suspect I could have felt much emotion for a goose. But this was not just any goose. This was *Peep*.

I called her by name. She stopped, lifted her head, and looked for a second or two, then casually turned and in a slow and graceful gait she walked toward me. As she got closer I heard the barely audible, throaty sounds that she always made when she was at ease and comfortable and "in contact." I hastily ran to fetch some lettuce from the fridge and tossed her some bits of it. She walked up to them, within four feet of me, reached down to examine the greens, and without taking a nibble or giving a nod of acknowledgment to me, she turned around and walked back to her gander who was watching at a distance, still tall, alert, and motionless. Once back with him, she stood up tall, stretched her wings giving them a few flaps, then slowly and deliberately tucked them back in over her back, briefly shook her tail, and bent down and acted as though she was picking at something by her feet, in what biologists call a displacement activity. I knew what she had just said: she had mixed emotions, but she indicated in no uncertain terms that, although she remembers me, she is now with *him*. She may have been imprinted on humans, but that had not prevented her from bonding with a gander.

They both stayed in the yard and she acted as though she had always been there. He remained tense. Three hours later, at 8:30 A.M., he started to honk, apparently to get her attention. Then he started rapidly nodding his head up and down while pointing in the direction of the pond. Even I could read his body language. It said: "I want to go!" She assented by responding with a few brief head-nods of her own and then *she* launched into flight. An instant later he, too, took off, first behind and then beside her as they flew down to the beaver pond.

I soon began to refer to the gander as Pop. Peep and Pop returned to our yard that same afternoon, and from that day forward the pair came twice daily, first at dawn and then again for a second visit in the afternoon or evening. By the second day, Pop already started to calm down, as my journal entry on March 30 indicates:

> By 7:20 A.M., they've stopped feeding and both are on the porch by the front door. She closes her eyes—I see the white of the tiny lid feathers as the lids close, and I also see her identifying scar on the upper left eyelid. Then she draws one leg completely up under a wing. She closes her eyes, but opens them again for a few seconds. The durations of her eyes staying closed become longer, and she tucks her head with closed eyes under the feathers on her back. He meanwhile stays next to her with his eyes open all the time. He stretches a wing, draws up one foot, and also tucks it under the

wing. Thus, they gently rock as they each balance on one leg, a picture of ease and contentment.

Pop at rest—one leg tucked up—on his favorite rock in the yard.

She soon falls fully asleep. He enters half asleep, perhaps literally, as he alternately opens his eyes for a few seconds and then closes them for a few seconds. Both birds can see me from 15 feet away, through the glass doors. Both feel comfortable perched on the porch; they've come here to *sleep!* They've both accepted us!

After half an hour she yawns, exposing her pink mouth lining, stretches a wing, and starts to walk down onto the lawn to feed on the cracked corn I've left them. He follows her, and after feeding they both return to sleep on the porch. This time she lies down before tucking her head into her back feathers. It's 8:20 A.M. An hour later he starts to make soft throaty grunts, then he holds up his head, faces toward the pond, and calls in synchrony with rapid up-and-down head wags giving his unmistakable message. She awakes, stands tall, and launches herself into flight. Within a split second of her takeoff, he lifts off as well, and they fly back down to the pond together.

Over the following days I noticed ever more details, and I became immersed in, if not mesmerized by them. I noticed that whenever Peep first started feeding, she continually made low, barely audible chattering—sort of a purr. Pop would answer at a lower frequency and lower pitch. Occasionally she would take the initiative to leave: As feeding continued, the volume of her purr would rise, as though she was getting more excited, and she raised her head and started to look off in the direction in which she would eventually fly, giving her rapid head-flick or head-shake (quite different from his signal—the up-down head wag). If he was feeding and didn't respond, she would bend down and take a few more bites, and try again later to rouse him. Failing that, she would just take off and he would follow. Pop never took off unilaterally. He always gained assent from her before the decision to leave was finalized, and the final decision was always hers. Once airborne, both would honk loudly.

Time to go.

Over the next two weeks I hoped the pair might stay to nest by the house, and to entice them to do so I nailed together a wooden platform on the lawn and topped it off with hay. But they never once perched on my construction. Instead, they always returned to the pond at night, even though the pond was hotly contested territory. The semitame resident pair were back, as usual, and other pairs of geese came on occasion as well. I saw vicious fights and in the violent frays I started to wonder who was fighting whom, and why. I suspected that the eventual winners would be the best fed, those with the most energy reserves. I tried to give Peep and Pop an edge by regularly feeding them cracked corn, both to shore up their strength for their fights with others at the pond and also to stimulate them to breed.

By this time I had, of course, been able to easily differentiate Peep from Pop, less by his size and generally erect bearing than by his unique angular white face bib, as opposed to her more rounded smooth one. I used my binoculars and started sketching the face markings of other geese to try to identify other individuals besides Peep and Pop.

Peep and Pop operated as a team. Invariably, when they came to the cracked corn, they approached as a pair and she fed first. She never showed hesitation to walk up to the grain and begin feeding

Takeoff—Peep leads in flight.

immediately. But only one of them fed at a time. When she fed, he stood tall and immobile nearby, head high, scanning all around. He never once barged in to feed ahead of her or with her. He would first scan in one direction, holding his head absolutely still for minutes at a time, then turning it slightly and looking in another direction, and so on all around. She, meanwhile, fed uninterruptedly and always without so much as a sideways glance. Only after she was sated, and indicated this by looking up, did he take a few steps forward to take his share of food. She then took over the watch. The same teamwork applied to sleeping. Both geese routinely took naps during the day, but never at the same time. As during feeding, first one would nap, then the other, in the same order as with feeding. And she did most of the napping that was done.

Peep had no fear of me whatsoever, but she acted as though I didn't exist. She had indeed briefly though pointedly acknowledged me on the morning when she first came back, but thereafter she seemed to obsessively make sure that all of the attention that she *used* to pay to me, was now conferred exclusively on her mate. I distinctly felt as though she was showing him that I was irrelevant to her, because when I was right next to her she *deliberately* looked away from me, rather than at me. Similarly, although Pop continued to watch me warily, he showed by his actions that he had

eyes only for her: Whenever he chased away another goose, he immediately flew right back to Peep and made a big show to her. He extended his neck toward her, lowering his head just over the water and doing much head dipping and shaking, all to an accompaniment of loud honking, in an impressive show that probably drew her attention to him. He was either taking credit in front of his mate for attacking a rival, or reassuring her that *he* was still there for her even though he had left her side for a half minute.

On March 31, Peep and Pop were late for the first time. They didn't show up at the house until the unusually late time of 7 A.M. Had they finally found their potential nest site? When they flew from the lawn to return to the pond, I followed them to find out. They had by then selected the upper smaller pond as their area of choice, possibly to avoid the intense competition of the long-resident pair who stayed mostly in the lower main pond. When I arrived at the upper pond, I saw Peep and Pop gesticulating to each other with necks extended. As they had done the previous day, they then paddled into a patch of sedge hummocks that was sticking out of the water near where the resident pair had nested on their first try four years ago. Somewhere in there, I felt, they had probably chosen a place to make their nest. But I return first to developments in the bog that led up to this point and gave me this premonition.

2 SETTLING IN

Within two days of Peep and Pop's arrival (March 28, 2000), other pairs of Canada geese had come to the bog and daily and nightly there were fights. I eventually called the pair that seemed to be the main defenders of the bog "the resident pair" because it was probably the same pair that had nested on the lodge for the two previous years. They were again perching on "their" beaver lodge and I expected them to start nesting soon.

The dawn of April 6 was cold, windy, and overcast. Ice had re-formed over half of the pond surface during the night. The ground was covered with snow. Peep and Pop, the resident pair, and also a third pair were present, and a lone goose was visiting. The three pairs were separated from each other while feeding, possibly on insect larvae, from the shallow pond bottom by periodically tipping over with their heads down and pointing their tails straight up. The resident pair were, as usual, next to their beaver lodge, and after they stopped feeding, they began a lively, water-thrashing bathing routine.

Bathing by geese is not merely a utilitarian behavior for cleaning the feathers, any more than kids' splashing around at the beach is for the purpose of washing the sand or dirt from their bodies. The geese's bathing behavior looks like play and they indulge in it most commonly immediately after repelling intruders or after feeding. Unlike perching birds, who bathe while standing up in shallow water and then skooching down and beating their wings to kick up a spray, geese about to bathe are already floating on deep water. They forcibly dip their head and neck deep down and then with curled neck they ladle up sheets of water but don't beat their wings. As this water runs over their backs, the birds reach over and rub their heads in it. When one of the pair starts bathing, its mate usually follows. After each series of water dips, the bathers rise up and rend the air with a few loud wing-flaps, some rapid tail-shakes, and a leisurely preening session.

As the resident pair were performing their exuberant morning toilette next to the lodge, I suddenly saw the gander stop his bathing routine and jump onto his mate's back to grab the back of her head with his bill. There was some splashing. It was a mating, which meant that eggs were now or would soon be laid. After their coupling, both came out of the water and walked up onto the beaver lodge where they again preened. The lone goose that had come onto the pond that morning gradually drifted toward them. When the loner got within about seventy-five feet, the pair flew off the lodge and both chased it. Peep and Pop were then some distance away from them, near me at the pond edge. Nevertheless, they flew up and joined the residents' chase after the lone goose. The third pair followed suit; all six geese had united to chase the lone goose off the pond. The resident pair, who either already had eggs or would start laying in a day or two, was by far the most aggressive of the three. This pair flew after the retreating goose even after it had left the pond. But they quickly returned to their perch on the beaver lodge. Curiously, I thought, Pop landed right next to them and was ignored. Three hours later all three pairs were still ignoring each other.

The next morning the resident pair was not on the beaver lodge, as I had expected. Instead, they were now by the muskrat lodge at the other end of the pond, possibly because the beaver lodge was being reoccupied by beavers this spring. The female was sitting down on this muskrat lodge (which was little more than a pile of decaying cattail leaves) possibly laying an egg while she was also reaching to the sides pulling loose leaves under and around herself. Her gander stood next to her.

Three days later, on April 10, their nest was buried under about a foot and a half of snow that had fallen in the night. Most of the pond had frozen over and no geese were near the nest. By 5:30 P.M. the next day, however, the goose was back on her nest, apparently having just returned and dug off the snow to uncover the recently laid eggs. Her fresh tracks showed that she had walked the twenty feet on snow-covered ice to get to the nest, and it was still snowing.

The snow had not let up by 8:05 P.M. when it was getting dark. Somewhere from up in the sky among the drifting snowflakes I heard the jubilant refrain of a woodcock's mating dance. The music seemed out of place now, as did the goose who, alone and immobile, was sunk down onto the nest.

It continued to snow all night and long into the morning, by which time the incubating goose was nearly buried. Incongruously, thick buds of sugar and red maples were ready to burst forth with flowers, and quaking aspen, speckled alders, and beaked hazelnut were already heavily tasseled in full bloom. Song sparrows, red-winged blackbirds, and phoebes had been back for two weeks although their songs had now become muted. The pair of phoebes that nest every year under our porch had spent the night fluffed out and pressed together into a single ball on their old nest next to the door. After arousing on this morning of cold whiteness, they silently perched at the edge of the woods with their wings drooped to the sides, typical of birds in distress.

The songbirds' activities resumed when temperatures soared (to 70°F) on the next day. But Peep and Pop still showed no signs of nesting.

As the sun started to illuminate the western shore of the pond on the morning of April 15, I finally saw a change in Peep's behavior that made me buoyant and hopeful that she and Pop might also nest. Peep swam to the east shore and climbed up out of the water onto a pile of debris among sedge hummocks. There she squatted down and started pulling dead vegetation toward herself while Pop stood by her with his head high, scanning around. Reproductive urges were awakening. Would they be realized? Canada geese normally don't nest until their third year. Peep was only two years old, but the extra rations of food I had been providing might coax her ahead of schedule. Only fifteen minutes after "playing" at nest-building with the sedge, she and Pop left the putative nest site, flew up to the house, fed on corn, and then went to sleep on a big rock by the driveway.

Over the subsequent days, Peep and Pop kept up their routine visits to the house. Neither returned to the sedge hummock that I had hoped she had chosen as a nest site. Instead, when they were not at the house, they loitered on an old earth-berm beaver dam in the upper portion of the bog. Unlike the residents, who were not even deflected by snowstorms, Peep and Pop seemed hesitant about nesting, and I feared they might soon miss their window of opportunity for the year. Finally, on April 21, after they left our house after their morning visit, I did not see them at the pond for most of the day. Did they search for another pond with less competition from other geese? In the evening I heard them call from afar, and then they flew over the house and descended to their old haunts, the small shallow pond at the upper end of the bog. At least there they could partially avoid the two pairs of geese in the main pond.

By May 1 the sugar maples were bursting forth in yellow bloom. Our phoebes chittered softly while flying to and from the new nest they were building under the porch. Peep and Pop came as usual to feed on cracked corn, and then to sleep while perched on "their" big rock at the edge of our driveway. But on this day they only stayed an hour, whereas up to now they had spent at least half of the day with us before returning to the bog for the night.

I went down to the bog in the evening and walked out onto the old beaver dam that bordered Peep and Pop's little upper pond. The Wilson's snipe that in the morning had been flying high over the bog whinnying in rising pitch in an unearthly sound produced by specialized "music" feathers, which act like the reeds of a wind instrument, was now making its slow, monotonous ticking vocalization while perched on the mud in a sedge-willow thicket. A swamp sparrow trilled and I heard the slow piping refrain of a white-throated sparrow, a bird that breeds far north of here in the conifers. This one must be on a brief stopover on his migration to the north woods.

Pop was perched uncharacteristically all by himself. Wondering what could have happened to Peep—they were never apart before for even so much as a few seconds—I searched and found her pressed down in incubation posture on a water-surrounded mound in the sedges. She had a nest! It was not the sight as such that made this scene so beautiful to me. It was the prospect of what it meant and what could follow. Against all odds, our Peep was settling in our bog with a mate, and they would try to start a family. I could not believe my good fortune—seeing her so close and yet in the wild, with a wild gander who no longer feared me. Furthermore, their nest was in knee-deep water and I could easily wade out to it and observe their home life.

3. AT THE NEST

Peep had her head up in a tall, graceful arch but dropped it onto the nest rim when she saw me. A goose's head, with the white face patch held aloft on its long neck, shows up like a beacon in the dark. Not surprisingly then, when Canada geese try to hide, they lay their head to the ground or the water. As I drew close, however, she suddenly pulled her head back up, looked at me briefly, and then assumed the sleep posture; she rested her head onto her back between the wings. She closed her eyes and made barely audible low grunting sounds when I walked up to her. If sounds have texture, these were velvet. I mimicked her sounds, to acknowledge them and our mutual contact. I felt a warmth of closeness with her. I knew from Peep's gestures that when she had first seen me coming through the woods she did not know who I was, but then when I came into the open she had suddenly recognized me and felt relieved! Pop, meanwhile, stayed on the beaver lodge, silent and relaxed.

I returned to the bog at midmorning and walked my usual rounds out onto the old beaver dam from where I could view her on the nest.

As I waded into the shallow pond to within two feet of the nest, Peep demonstrated her trust in me by closing her eyes and going to sleep, while Pop watched without showing any reaction. I sat down on a sedge hummock next to the

The ancient upper dam that I walked on daily to get to Peep and Pop's nest.

nest and feigned indifference toward the geese and their nest, paying attention instead to a male red-winged blackbird who hovered over me and fluffed out his scarlet epaulets framed in yellow. He was scolding me because at least one female of his species had a nest close by that was sunk into the center of another water-surrounded sedge hummock. Unlike the male-female look-alikes of monogamous geese who chase either sex off their bog, the brashly colored redwing males presumably entice any number of their drab-colored females to settle in their territory. But enhanced

male showiness does not necessarily indicate polygyny. I watched two pairs of goldfinches—the males were already in all their flashy yellow finery in contrast to their females' pastels—noisily circling the bog. A yellow warbler just back from migration was singing in the willows that grew bushy where the beavers had pruned them. Two male wood ducks and one female swam along the farther shore and were monotonously calling *katik, katik, katik.* None of these three species with brilliant males defends territories to keep harems, as redwings reputedly do.

Pop did not stir except to occasionally look toward me and Peep. After about half an hour I stood up with the intention of trying to touch Peep on the nest, but I had taken only one step toward her when Pop started swimming toward me in what looked to me like a deliberate but unhurried manner. I stood still. He continued to paddle up an open channel of water in the sedges, and then when he got within six feet of me he hissed and without as much as a moment's hesitation flew up out of the water directly at my head. I raised my hands to ward off his powerful wing-strikes. I stood still, but he came at me again. After his third attack I slowly backed away from the nest. His loss of fear of me was not such a good thing, after all, unless I also taught him to trust me.

On my next visit I brought along a mugful of cracked corn to appease him. Normally, Peep did not leave the nest, even for corn, because she took her feeding breaks on her own schedule. This time, she hopped off the nest and hurriedly fed on greens next to me and within ten feet of the nest. She pulled up newly sprouting sedge blades to eat the lower, pale white parts after neatly snipping off the harder, darker tops using the serrated edge of her bill like a cutting knife.

Standing next to the nest, I saw for the first time her four huge yellowish eggs. As I had anticipated, Pop came and hissed at me. I stood my ground and threw grain onto the water in front of him. He stabbed at it viciously, alternately feeding and hissing, as though attacking it rather than me. Peep took grain gently out of my hand as she paddled around my feet.

I sat down on a sedge hummock next to the nest. Pop again lunged at me. His attack was not nearly as vigorous as the previous day's and I turned my head and continued to hold my ground. He seemed to calm down, but he watched me continuously as Peep fed on the few remaining corn kernels in the water, until a muskrat popped its head up out of the water near us. Pop spread his wings and jumped at the retreating rat. Thirty-nine minutes later Peep returned to the nest, preened, and settled herself down onto the eggs while rocking gently from side to side. She reached around with her bill and pulled nest material tight all around her. Ten minutes later she stood up briefly and rolled the eggs with her bill before again resuming her incubation. I was surprised that she spent so much time off her eggs, but I learned later that negligence could be the norm in a goose's first nesting attempt.

During this time the resident pair from the main, adjoining, pond had honked briefly. Pop instantly dropped down into hiding posture. Why was he hiding from them? Peep, instead, raised her head attentively, then put it back down and slept. She kept one eye shut while the other remained open.

May 3, 2000.

Pop is now tolerant of me and I like being with him at the nest. He doesn't beat or bite me anymore, and he also ignores beavers that come close by. I'm beginning to wonder if the geese see *themselves* as their worst enemies: Two geese have just flown onto the adjacent pond where the resident goose is now incubating her eggs on the muskrat lodge. A big commotion ensues, as the resident gander chases these geese all over their pond.

Early in the season Pop had always joined in the fray to help the resident pair, but now he no longer offers his neighbors assistance. Instead, he ducks down behind some sedges and lays his head horizontal with the water so as to hide himself. Why does he now hide whereas he had always been noisy and defensive before? Shouldn't having a nest and eggs make him more defensive rather than less?

Pop's behavior toward me was, until he got to know me, typical of ganders. When I had approached the nest of the resident pair, the gander immediately jumped onto the nest next to his goose and hissed at me while standing tall and spreading his wings. I did not disturb him further. I instead went on to probe another gander's reaction at another pond where the geese did not know me at all. There the gander also became increasingly excited and threatening the closer I came, and attacked when I was still ten feet from the nest by flying directly at my head. I was thoroughly intimidated and did not dare to come closer. Unlike Pop, this gander had not been tamed, but his defensiveness was strong enough to overcome his cautions.

May 6, 2000.

Ruby-throated hummingbirds, rose-breasted grosbeaks, and crested flycatchers have returned. I'm happy to see spring on the way, but the upper beaver pond is nearly drained. The water moat around Peep and Pop's nest is almost gone. Below the dam I see a beaver who is probably looking for a leak, and I walk the length of the dam looking for it myself. Finally I find that water is seeping out through the bottom. Will the beavers solve this problem? The quiet environmental crisis is often the most profound.

Peep stands up on the nest when I return to it, and as though on signal, she gets up and covers her eggs by pulling her recently shed down feathers from the periphery of the nest over them, and then she flies off to land in the water fifty feet away and immediately starts dipping for food. Normally Pop joins her when she leaves the nest. But now he stays next to me, and he hisses at me in agitation. He is torn between joining her and staying to defend the nest. It's *his* problem, but he blames me for his dilemma. Peep swims back to me and the nest in a few minutes. She strains mud to find food before climbing back on and settling in to incubate. I stroke her back with my hand and pull up dead sedges for her. She takes sedge leaves from my hand and tucks them under her around the nest edge. Pop is at ease only five feet from us. Once more Peep hops off the nest. The eggs are unprotected, but Pop still seems unconcerned even as I touch them. Peep swims away, and Pop again becomes belligerent toward me. He relaxes as soon as Peep comes back toward me and the nest, making soft grunting sounds that show he is at ease. Peep stands on the nest and preens for four minutes before settling onto the eggs. After several more minutes she stands up again, rolls the eggs with her bill, lies down again, and tucks more nest lining in around herself. Pop remains calmly by the nest as I stand up from my sedge seat, wade back to the beaver dam, and walk out of the bog.

May 14, 2000.

I've meanwhile paid little attention to the geese on the adjacent larger pond where the resident pair has been incubating their six eggs on the muskrat lodge for the last month. But today a long silence is shattered when another pair, which has been tolerated by the residents and has not nested, makes a big commotion. The resident pair remain silent, as though they have nothing to defend. Then, I see them with four goslings that have apparently just hatched. The family is swimming west, directly across the middle of the pond. [I later found two rotten eggs left behind in the nest.]

May 16, 2000.

I looked in vain for the geese and their four goslings. Thinking the family might be hiding, I began a through search. I waded through all the sedges and cattails. I explored all over the pond and around it. There was still no sign of them. I was forced to conclude that the "residents" had yet again left the pond, as in the previous three springs.

The pair that now remained on the pond did not nest, though after the residents and their goslings left, they examined their former nest site. Perhaps they had also wanted to nest at that spot and now saw their chance. Unfortunately the window of opportunity was past. It was too late in the season to finish a nesting cycle.

Although I found no goslings or their parents, within an hour of searching I stumbled on two new red-winged blackbird nests with eggs, one song sparrow nest with eggs, and I discovered a rose-breasted grosbeak starting to arrange the first three to four twigs of a nest platform in a *Viburnum* bush. The song sparrow, instead of flying off her nest, ran through the grass convincingly mimicking a mouse.

Six male goldfinches in their flashy yellow-and-black nuptial plumage alternately perched on trees and chased each other over the bog. Some females tagged along. The birds circled high in small groups that separated and again joined. Goldfinches nest in the stunted isolated trees in this bog every year, but not until at least three months from now when thistles start to bloom on the farmers' pastures that have seeds in time for the young. The pairs are courting already and apparently establishing nesting territories months in advance.

May 18, 2000.

For the hour after sunrise (but not afterward), the male red-winged blackbirds were frenzied. All the females that I happened to see flew fast and acrobatically, usually into thick bushes and trees where they tried to shake male pursuers that usually numbered three or four and up to six at once. The males—contrary to the map diagrams in scientific publications on red-winged blackbirds I have seen where "territories" are depicted in nonoverlapping dashed lines—did not stay within designated boundaries. Instead, they routinely flew from one end of the bog to another. Two males commonly perched within two to six feet of each other, acting oblivious to one another. In one hour I saw not one male chasing off another male, and many males flew all around. On the other hand, I saw 75 chases of males behind females, and not one led to a copulation at the end of the chase. Even though it is the height of the egg-laying period, a casual search revealed just six fresh redwing nests. By around 7:30 A.M. the males inexplicably stopped their chasing and they ignored vocal and conspicuous females next to them.

May 19, 2000.

I heard the first tree frog yesterday—the sound of summer. Then, last night, there was frost. Now I'm at the pond, watching the sun rise and seeing it burn off a layer of fog hanging over the water. The bushes and the grass are laden with glistening water droplets. A cool breeze riffles the pond, and a deer along the opposite shore snorts several times. I pause to map in my mind what birds (aside from the grackles and redwings) are singing and where, locating two willow flycatchers, three song sparrows, three swamp sparrows, two yellowthroat warblers, two chestnut-sided warblers, one yellow warbler, and a northern oriole.

I walk onto the beaver dam that I use every day to get close to the nest and Peep cranes her neck as she sees me coming. Pop, who floats on the water nearby, also raises his head and then drifts toward me. At 6:31 A.M. she leaves the nest and swims directly toward me. He follows her. She grazes on grass at my feet, drinks, and preens. He is within four feet of me, occasionally nibbling on a grass blade, but generally watching in different directions and only seldom glancing at me. Meanwhile, in the main pond beyond the old overgrown beaver dam, the nonnesting pair of geese continually look over toward us. They paddle aimlessly over the pond and are noisy. They seem to be excited.

At 6:36 A.M. the sun peeks over the hill, and a few minutes later Peep swims back and climbs onto her nest. Pop accompanies her to the nest edge, making soft humming sounds, and after she settles down to incubate he again drifts back over to me. He stays close and at times even follows me as though we are allies in the prolonged watch of Peep on her nest. We then both stand still, attentively watching and looking in all directions and listening. A pickerel frog sounds just once to the left, and to my right in the woods a hairy woodpecker drums repeatedly.

Pop stays by me as if he were keeping me company. I am amazed at his apparent attachment to me because, unlike Peep, he grew up wild. Does he sense that I can help protect them from danger? Does it make him feel at ease to see me scanning the skies and the bog? He remains silent even as the other pair continues to be noisy, animated, and provoked. I can find rationalizations for Pop's behavior, but the nonbreeding pair leaves me baffled. They seem aggressive, but have nothing tangible to defend.

I remain standing on the dam next to Pop with Peep just beyond on the nest. I watch the female rose-breasted grosbeak as she picks twigs and makes several trips to her nest nearby in a *Viburnum* bush. She is always accompanied by her mate, who is now silent. A second male stays nearby for about a half hour, but surprisingly the first male does not chase it. The second male finally sings, but only briefly as it flies away. His wing-beats are slow and deliberate. They look more like the displays that redwing males give when they hover over the bog, than the flight pattern used for covering distance—a series of a few rapid wing-beats alternated with short dives when the wings are held close to the body. Still a third male rose-breasted grosbeak sings loudly and continually from the other end of the bog. The grosbeak's languorous, harmonic song is one of the most enchanting in the bird world. I never tire of listening to it, maybe in great part because it is not intended for me. It is meant as a territorial claim for other males and to attract a mate. Not surprisingly, many birds stop singing once they have a mate.

One song sparrow near me continually sings from perches above the low vegetation. Another one occasionally chimes in. Three willow flycatchers, who have only recently arrived from migration, call from different directions. I can also sort out five different singing male yellow warblers. All of this music to my ears represents ongoing social flux at the beginning of the breeding season. I enjoy it now, knowing it will end soon. The swamp sparrows and the yellowthroat warblers are already almost silent.

By 7:40 A.M. the fog is gone and the first barn swallows start to skim over the pond. A pair of tree swallows land on a dry dead elm that stands alone by the water's edge where a kingfisher often waits to dive. As I leave for breakfast, I hear a single, quick refrain of the high-pitched shrill of a blackpoll warbler. This bird breeds up north in the conifers; migrants are still coming through.

Most leaf buds (except the always tardy American ash) are opening or opened, while seeds of the quaking aspen have already ripened and float through the air like snowflakes. Tassels of oak flowers have dried and are ready to fall. Apple trees are starting to bloom. The rhythm of the bog is unfolding just as quickly and within it that of the geese.

This morning the redwings behaved astoundingly different from yesterday. Today the males reacted to each other. Shortly after dawn (when blackbird activity peaks) almost all the interactions were between pairs of males that spiraled about two hundred feet up into the air above the bog. These male-male pairs flew in unison as if performing some sort of choreographed ritual dance, staying within about five feet of each other. Looping lazily, they made no attempts to close the distance between them or to evade one another. Instead, they flew with slow exaggerated wingbeats, then drifted down with spread tail and wings, and accompanied that display with the *oog-la-eee* song. After as long as five minutes in the air, the two or three displaying males descended into the brush, usually to be joined by two or three more.

Over the next several days, as I continued to visit the goose nest, Pop started to act cautiously and suspiciously. He occasionally hid while I was approaching the pond through the woods, although as soon as I entered the open bog and he could see me plainly he relaxed and came toward me. He could see me from a distance, but he may not have recognized who I was. I did not do proper experiments to examine his powers of individual recognition. However, I did make a probe. One afternoon I came with my teenage son, Stuart, who stood back as I went to the nest and patted Peep on the back and then reached under her to feel her surprisingly warm feet. I took an egg and held it up for him to see. Peep ignored me. Pop watched from a distance, also unconcerned. Apparently he now trusted me totally. Would he trust someone else? I left the nest, and Stuart then waded toward it. Pop immediately swam over to intercept him and Peep also became agitated. Stuart, eyeing the big hissing gander approaching, unilaterally decided that the experiment was conclusive enough.

Later that day I asked another teenage boy (also my height, build, and hair color) to approach the nest. The results were the same: Pop hissed threateningly when John came within fifteen feet of the nest, and when he took a couple more steps Pop struck him in the leg with his bill. Peep rose up in her nest and also hissed. After the rout, Pop rose up tall and beat his wings in a gesture I took to mean that he was at least as satisfied with the end of this experiment as we were. However, the unintended and hugely unanticipated travails of my geese at this nest had scarcely begun.

4. NESTING TRAVAILS

It is a cool, overcast dawn at the bog (on May 22, 2000). The male red-winged blackbirds are noisy in the cattails as usual. Two barn swallows skim like erratic winged arrows over the pond surface as a beaver circles its lodge and then crawls up onto the dam where it snips off fresh willow shoots. Only the geese seem agitated.

I hear their cries long before I get up. Shortly after I reach the bog, the pair of nonbreeders, who now have exclusive use of the large lower pond, continue to call noisily. They fly up and circle above Peep and Pop (and me) at the upper pond and then return to their home area. Peep and Pop reciprocate: Making agitated calls, they leave from their nest and fly over to them. A huge commotion ensues and since Pop used to *hide* from them, his going out of his way to meet them at their pond is strange, if not bizarre. It is presumably not without reason. No goose behavior is.

I leave the dam, go over to the lower pond, and watch the continuing chases and standoffs, hoping to uncover some clues. I see a swallow pick up a goose feather that is floating on the water, fly high into the air, and then drop it. As the feather drifts down, the bird catches it, drops it again, and repeats the game several more times before finally flying off with it in its bill. For the time being, I'm not much wiser, but I suspect the loose feather is a result of a fight between the two pairs.

When I return in the afternoon, the geese seem at ease. I listen to the rose-breasted grosbeak. He seems to be singing directly from a just-finished nest. Since it seems odd for a male to sing directly by a nest, I go over to check if maybe his mate is sitting on eggs. To my astonishment *he* is sitting on the nest himself, although only the female in this species incubates. As I look closer to reassure myself that it is really the male, I see proof—his bright crimson breast framed in immaculate white. And he is not merely perched on the nest rim. He is firmly sunk into the nest mold of dark rootlets. Seeing me, he rises and flies off. His drab-colored mate is not around, but she must have been there recently because the nest contains an egg. It is only *one* (of the eventual four) eggs of a rich, deep turquoise background blotched with russet-brown. I do not know why the eggs are so richly colored, why he is sitting on the first egg, or why he sings while doing so. But I have some ideas.

The baby robins hatched today from the azure eggs in the hard, mud-cup nest lined with sprigs of soft, straw-yellow grass that sits in a low fork of a speckled alder at the bog's edge. Nearby in a twiggy low spiraea bush a yellow warbler is sunk into her nest of cattail down. The nest is so deep that the female disappears from view into the pale, solidly felted fluff. Only her tail and bill stick straight up. The cattail down with attached seeds was produced last summer, and is only now becoming airborne and used by yellow warblers to line their nests.

May 23, 2000.

I stop at the song sparrow nest just before step-ping onto the beaver dam. Last night the incubating female slipped off her four eggs and ran through the grass like a mouse. This morning she stays put. I look closer, and closer, and finally see—she is *dead!* I pick her up and find not a mark on her. She expired in the night while incubating her eggs. A *Necrophorus* burying beetle will soon find her, "call" for a mate by broadcasting an alluring scent, and together they will bury the bird and reincarnate her into the next generation of these handsome orange-and-black animals. There are necessarily many individual deaths in the bog, although they largely go unnoticed because they are soon recycled into other life in an unending chain.

As I walk onto the dam (it's nearly 6:00 A.M.), Peep and Pop both come to me immediately. Until now she has always fed as though in a frenzy. Yet today they amble around like zombies, only nibbling on the grain and occasionally picking at some greens or stripping a sedge flower. (Instead of biting off the sedge inflorescences, geese hold them loosely in the bill and then with an audible *rip* they jerk their head upward, leaving the central tough flower rachis on the plant and ending up with the tender flower parts in the mouth.) They have always rapidly and methodically stripped one flower after the other. Not today. Today they mouth only a few lackadaisically.

Isn't she going to return to the nest, I wonder? I wait, and wait, and walk idly to check on nearby redwing nests in the sedge hummocks. Peep follows me! That's absolutely absurd. She has never done that in Pop's presence. From the moment we remet this spring, she has gone out of her way to turn her back on me when her mate was near. And now all of a sudden she shows more interest in me than in him. It might be a coincidence, so I test her by walking on the dam. She still follows. I'm baffled.

Pop seems to be getting tense. Finally he has had enough. He starts honking and head-bobbing, telling her "Let's go." In the past he has never flown off until Peep has indicated that she will come, too. And she usually gives her assent quickly. But not today. Instead, she ignores him. He continues his solicitations for minutes. She takes no notice even as he comes up close beside her. Finally she leaves him altogether and paddles aimlessly among the sedge tussocks.

I remain on the dam, and she eventually swims back to me, and then pointedly *looks up* at me. This is getting to be bizarre! Until now she has always looked away from me if I looked at her directly, or she tucked her head into her back feathers to sleep or feign sleep as if to avoid my eyes. I bend down and look directly into her eyes as if to ask rhetorically "What's up, Peep?" Does she turn her head? No. She tweaks my nose with her bill! This is more strange than I can believe, and I would not believe it now if I had not written it down then.

An hour passes and Pop again entreats her to leave with him. This time she finally responds. But inexplicably they do not fly to the nest. Instead she leads him in the opposite direction and they descend into the main pond. Instantly I hear a loud medley of honking and the violent thrashing of wings in water in a bay around a wooded peninsula. I must run through the woods to get a view of them: Peep and Pop are crowded next to each other near shore and the nonbreeder pair, about fifty feet from them, is slowly closing in. Pop gesticulates wildly with head-bobbing and honking. He wants to leave.

The other pair is blocking the bay as if to corner Peep and Pop. The standoff continues for about half an hour. Peep and Pop eventually "escape" the blockade by forcing their way sideways through dense sedges at the edge of the pond. The nonbreeders start to pursue but stop about fifty feet from them. The two pairs engage in a loud honking exchange. Pop gesticulates wildly, bowing his head, keeping his long neck outstretched and his bill open. He again incites Peep to leave, but she remains unresponsive, and so he starts to jab her in the side with his bill for emphasis, something I have not seen before. Finally, at 7:28 A.M., nearly an hour and a half after I first saw Peep and Pop this morning, she shakes her head in the gesture that says she is ready, and they both lift off and fly. Very uncharacteristically, *Pop* takes the lead in flight. I had, of course, expected them to cross the hundred or so feet of sedges and cattails to fly back into *their* pond. Instead, they fly up to our house and there continue their calling. This is their first visit up to the house since she started incubation.

Something is seriously wrong and I presume that it concerns the nest. So instead of going to them at the house, I run back to the upper pond, strip off my boots, and wade out to check. As expected, there is a major problem: Three eggs are lying in the mud on the pond bottom. The one egg remaining in the nest is slightly warm, so the disturbance that caused the birds to leave must have been recent. I had probably just missed it by a few minutes when I first came down this morning.

Not knowing what to do, I pull the ice-cold eggs out of the water and put them back into the nest, even though the developing embryos are likely dead. For the eggs to have even a slight chance of hatching, Peep has to get back onto the nest immediately. So I sprint through the shallow water to and along the beaver dam, through the bog, and up the driveway and to the house to chase Peep and Pop into flight. I hurry back out to the nest again, and they follow. I'm brusque and aggressive, which induces Pop to defend the nest. Peep hops onto her nest, preens for two to three minutes, then settles down to incubate. *Great!* I've salvaged what I can, and I leave. Seconds later—I've barely waded back to shore—she stands up and fusses with the nest and eggs with her bill. Then she leaves the nest! Maybe it has all been for naught.

Meanwhile, the nonbreeder pair is again making a huge racket in the neighboring pond. Their heads are held high and they watch us. Curiously, Peep and Pop remain totally silent this time. I expect these wild geese to fly over and attack any second but my presence probably inhibits them. What are they up to? I stand still trying to be unobtrusive, then climb a tree at the edge of the bog to view both ponds and watch what develops. At 9:30 A.M., the others finally fly off in a westerly direction where they have often flown in the past.

Peep and Pop now return to the nest once more. This time *both* climb onto it, but neither squats down to incubate. Peep soon leaves and Pop remains standing on the nest. He stands on it for fifteen minutes and pokes with his bill and occasionally rolls the eggs as though inspecting them. I have never before seen him or any other gander touching the eggs. It appears that he senses something amiss, that something has to be done with the eggs, but he does not know what to do.

Peep swims around in the channels through the sedges near the nest as I watch from my seclusion in the tree. Pop finally abandons the nest and follows her. I'm surprised to see her paddle to the north end of the dam and pause there at the precise spot where I daily appear and disappear on my way to and from them at the nest. I have never seen her there before. She raises her head high and looks long and intently, again and again, into the woods along the path where I travel. Finally she leaves the water and, walking *into the woods* where I usually go, she stops every few feet with head high, looks ahead, and advances some more. If she is looking for me, then it would be extraordinary behavior, and acceptance of exceptional phenomena demands vigorous tests and special caution. Her behavior *seems* to be self-evident, but there is still a possibility that she is hungry and looking for the grain that I carry. I must test if she wants me, or corn.

After she returns to the water, I rush down from the tree and race home as fast as I can to get some corn. I come back and leave a pile of it on the dam. As expected, she comes to me, but she takes only a few kernels, then follows me as soon as I walk away, leaving the corn. So if it's not food (I have not fed her anything else), why does she follow me? It's a mystery, and I'm pleased to think that there is more depth to this goose than I had dared to imagine.

After returning to my study to write up my notes and thinking about what I had just seen, I wondered if Peep somehow "knew" (perhaps unconsciously as expressed in a programmed reflex) that, since the eggs were cold, they were no longer worth incubating (i.e., dead). If she rejects eggs because they are cold, then she should accept *warm* eggs! Bingo! The thought is followed by my jumping up from my desk and running down to the pond with a bucket of warm water. It's now 10:20 A.M. She is on the old beaver dam with Pop at their favorite hangout. I rush out to the nest, grab the eggs, bring them to shore on the dam, and dunk them into the bucket. A few minutes later I put two of the rewarmed eggs into my coat pockets, and with one in each hand I wade back out to the nest. As before, my choppy actions probably seem aggressive or strange to Pop, who intercepts me and strikes me twice with his wings. But I manage to return the now-warmed eggs into the nest. She comes, climbs onto the nest, and preens as she always does before incubation. But she preens for only three minutes. Then she sits down, rocks slightly from side to side, and positions herself to incubate as though nothing were amiss. She closes both eyes and preens her back while both eyes still remain closed. Usually she keeps her eyes open while preening. Maybe she is tired after a rough night. Within one minute she tucks her head onto her back feathers and goes to sleep. She has resumed her incubation.

It is pouring rain when I arrive at dawn the next day. Peep is still incubating, and the nonbreeding pair is in the middle of their pond, bathing in the downpour by ducking down with their heads and throwing sheets of pond water onto their backs. I wonder what they have done to feel so proud of themselves, and seeing them so contented and having suspicions about their agenda, I feel uneasy.

I come back to Peep and Pop's nest again in the late afternoon when the sun is shining. Peep immediately gets up, covers her eggs by pulling nest material from the side over them and then leaves the nest to feed on grass on the dam. Pop accompanies her and he looks frequently toward the other pair on the other pond.

A muskrat pops up out of the water near us at the edge of the dam. It swims at the water surface into the tussock channels. Despite being in plain view, Pop gives it only a passing glance, continuing to look toward the geese on the other pond; Pop's behavior unequivocally shows me that muskrats are innocent of the nest attack. In any case, why should a muskrat roll out eggs? And how could a rat get close to the nest with Pop defending it? Many lines of circumstantial evidence point to other geese being responsible for the nest attack, although I can't yet think of a good reason for it.

Unlike yesterday, Peep does not come to me. After thirteen minutes she returns to the nest, stands over the eggs and preens, and two and a half minutes later settles down and resumes incubating.

Rachel, who is a behavioral biologist, had been as puzzled by my recent frantic activities as I had been by the geese's. Why would a man want to run down the driveway carrying a bucket full of water to the pond? She was (like me, I think) on the right track with regard to the geese. In my explanations about the water-carrying I also told her of Peep and Pop's confrontations with the other pair and how Peep inexplicably sought me out. To this Rachel remarked, "Maybe they won because of you—you came and then the other pair stayed away. Maybe that explains Peep seeking you out. She needed your help. It's a lot to attribute to a goose, but it is consistent with the facts." Accounts of animals developing trust in humans and counting on them for aid are not unprecedented. At Shark Bay, Australia, for example, an unknown wild dolphin with a hook in a badly infected jaw approached the humans who were there studying free-living dolphins in the bay. This animal came up to a researcher, and held still while the hook was removed from its jaw (Personal Communication, Rachel Smolker). Closer to home, a neighbor one night intervened as a raccoon was in the process of trying to kill one of her domestic geese. The raccoon ran off and the goose with its lacerated neck ran up to her and pressed against her, which the goose had never done before. Like the dolphin, this goose did not struggle even as it was being handled and (successfully) treated. Similarly, a raven with its feet badly tangled up with musk ox fur approached two kayakers, Lonnie Dupre and Tom Hoelscher, in the wilds of Greenland after they pulled up to make camp. They spoke softly to it. The raven picked up a small rock, looked at them, and put it back down. They then also picked up the same rock, put it down, then scooped up the raven, untangled the musk ox fur from its feet, and threw the raven into the air. It circled, landed on a rock, squawked, and then flew off to a cliff from where it had come (Personal Communication, Lonnie Dupre). Such anecdotes should neither be uncritically accepted, nor should they be dismissed. They simply leave me more puzzled than before.

My geese provide me with more questions every day. If I hope to answer at least some of them, I need to observe many more intimate details under all sorts of circumstances and try to see how they line up. Right now a lot of the facts are not lining up into any consistent pattern, and that is making the geese more fascinating.

My intuitions suggest that the nest was disturbed by the other pair. But so far my logic is inadequate to explain why a pair who didn't even bother to breed would invest time and effort to attack others, and ultimately risk injury in their fights to destroy Peep and Pop's nest. The nonbreeders had lived peaceably with the *resident* pair on the same pond, and that other pair *did* hatch young. Coming all the way over to the other pond to attack Peep or her nest is at odds with my current standards of rationality. Still, regardless of whether there is an adequate reason, the detailed observations converged in an intuition. I suspected that pair, and feared (predicted?) that they might do it again.

While I'm having breakfast in the kitchen on the morning of the next day, May 26, I'm suddenly surprised to hear Peep and Pop calling and to see them flying around the house. They land on the driveway. Something has indeed again happened at the nest. I drop my toast and rush down to inspect.

Three eggs are slightly warm and covered in the nest. The fourth is under water nearby. The goose almost always covers her eggs before leaving the nest, so Peep had not left in haste as during a predator attack. Might she have ejected the egg herself? I doubted it, but possibly they eject dead eggs. So I took the ejected egg home and cracked it to find that the fetus inside was indeed dead. However, if it was killed by the long ice-water immersion three days ago, then the other eggs that had been submerged (one was not) would also have been dead. If Peep could discriminate and then selectively eject dead eggs, why did she not throw all the eggs out except the one that had escaped the icy cooling? No. It didn't make sense. If, as I had recently determined, even ravens accept and incubate rocks, flashlight batteries, and potatoes, then geese would not be more discerning and purposefully eject eggs with a recently dead embryo. It is clear: The eggs were dislodged during another row, and since Peep had taken time to cover the eggs, it probably occurred after she had left the nest. It also seemed likely that whatever was attacking the nest did not do so in order to eat the eggs.

Next morning, I awaken a little before 6 A.M. to the racket of geese in the bog. I rush down and find Peep on the nest. Pop is nearby, alert but silent. Both are looking in the direction of the other pair who are calling loudly and swimming back and forth at the near end of their pond closest to Peep and Pop's pond. Their attention is directed at Peep and Pop. After a while one of these nonbreeders flies up, circles directly over us, and then returns to its own pond. Pop then starts calling and energetically gesticulating "Let's go." Peep answers, rises off the nest, and hastily covers her three remaining eggs; then they both fly over to the other pair for a violent and quick attack, after which the two pairs separate and regroup, noisily display to each other, and then bathe. They stay in tight pairs at the lower pond and continue loud outcries in what looks like a shouting match, for seventeen minutes, before Peep and Pop fly back to their nest.

This time Pop does not act aggressively toward me when I reach under Peep to pull out one very warm egg at a time. I examine them closely, wondering anew about the outside chance that she might throw out her own defective eggs. I'm convinced she won't, but I test the theory anyway. I offer her a smooth granite rock and a kiwi fruit. She turns both with her bill in the slow deliberate manner that she handles her own eggs. She manipulates both objects indiscriminately with her own eggs, and then settles down to incubate them, making her little comfort sounds that I love to hear. (I later remove the fake "eggs," knowing she will continue incubating her own eggs regardless of their contents.) As I listen to her comfort sounds, I feel solicitous and offer her a few handfuls of dry sedge. She grabs these inedibles and immediately tucks them all around her; she's still "making" her nest even though hatching time is close. That behavior is now no more necessary than the incubation of dead eggs. Both are, however, consequences of genetically coded behavior that *is* essential.

Two red-winged blackbird females have their nests sunk into sedge hummocks close to Peep's nest, and today their naked pink hatchlings emerged from the blue eggs with lilac and black squiggles. The male is defensive, circling over me. Peep keeps cocking her head and watching him, too.

May 28, 2000.

The first cedar waxwing pairs are only now returning to the bog, and within days they will start to build their nests of green moss, dead grass, and lichens in the *Viburnum* bushes growing on the old beaver dam. I walk around the bog to get acquainted with the other birds. In just half an hour I find three grackle nests in a loose colony within two to three inches of the water in last year's dry cattail leaves where the new green shoots are just now coming up through the mud. Two of the nests have five eggs; one has four. The yellow warbler nest has four eggs, a song sparrow's has five, and besides the red-winged blackbird nests with small young, another redwing nest contains two just-laid eggs. Another is freshly built and ready for eggs. The rose-breasted grosbeak nest was robbed four days ago, but the pair already has a new half-finished nest and will start laying a new clutch in two to three days.

As I stop to see Peep after finishing my bog inspection, she leaves her nest, swims up to me on the beaver dam, and rises tall and flaps her wings. I reach down to touch her feet—I have noted they feel hot when she incubates. She shakes her leg in mild irritation as I touch her toes. They are cold. She jumps into the water and swims to the nest. I follow her, and as she waddles onto the nest and stands to preen I feel her feet again, reaffirming that they are now warm to the touch, despite their recent immersion in the cold water. Thus, she prewarms them in *anticipation* of incubation. She soon sits down onto the still-covered eggs, rocks her body from side to side, and kicks the covering of down feathers and dried vegetation off the eggs with her feet. I again reach under her to feel her now very warm feet on top of the eggs. She doesn't struggle even as I cup my hands around her folded-in wings and pick up her entire body and set her back down again. Pop is right next to us, and even he is at ease. He would, of course, not tolerate me picking him up, but if there is any other way that this wild-grown goose could show more trust, then I don't know of it.

Two days later, on May 30, after Peep had been off the nest and with Pop for only ten minutes, Pop took the (unusual) initiative to leave her and go to the nest himself. She did not follow, but while there he kept shaking his head and looking to the west, as though wanting to fly in that direction. Why would he want to *lead her off* with the eggs presumably now almost ready to hatch? She continued feeding, and Pop then left the nest. Eight minutes later he pointedly returned to the nest once more, again standing on it, this time for twelve minutes. All the while he kept watching her, first as she stood on the dam by me, then as she paddled through the sedge labyrinth, and when she returned once more to the beaver lodge near the nest. He made loud purring noises and kept looking down at the nest, and when she came near him, he emphatically wagged his head up and down to again try to lead her off to the west.

She seemed nervous and distracted and finally started head-shaking as well, indicating that she was ready to follow. Were they considering something that had to do with the expected hatching? He nodded his head in a quick response, and then as she flew off, he followed. They flew first into the neighboring pond; then they flew up again and continued on. Knowing they would soon be back at the bog, I climbed high enough into my lookout tree—the sugar maple at the edge of the bog—to watch both ponds at the same time. (At this time I had no clue about their curious behavior—it would only be revealed to me with other pairs later.) I did not think about why they left, but I wanted to find out what would happen when they returned.

It is pleasant among the fresh green foliage, with a good view to the pond. The sun is bright, but despite the pleasant temperature, I hear very few frogs: a tiny peep from a spring peeper (they are mostly gone by now), one croak from a green frog, one purr from a leopard frog, and one brief series of calls from a tree frog. In contrast, the birds are vocal everywhere and all at once. Down below in the cattails the redwings predominate. But I also hear three rose-breasted grosbeaks at three different locations around the bog. I hear an oriole, yellow warblers and yellowthroats, redstarts, swamp and song sparrows, goldfinches, willow flycatchers, blue jays, a black-billed cuckoo, crested flycatcher, and grackles. A pair of redstarts hop close to me, and I see the female snatch and swallow a caterpillar. A pair of robins zip by noisily just below my feet. Two kingbirds perch on dry willow twigs next to the beaver lodge, and they, and several cedar waxwings, make frequent sallies high up into the air to hawk insects emerging from the bog; then they almost float back down on spread, motionless wings. A male ruby-throated hummingbird stops to hover near me, and down, close to the water, a marsh harrier sails by. Finally I hear the expected gaggle of geese returning. Almost immediately the noisy whistling of wings follows and Peep and Pop are rapidly dropping out of the sky on outstretched wings. For the moment I don't wonder about why they left or where they have been. I'm more enthralled to see them descend in unison in a graceful glide onto their pond and ski to a stop on outstretched, forward-directed feet, as Peep then paddles forward and hops onto the nest. Pop follows her to the nest edge. She perches on the nest for two and a half minutes before settling down to incubate, and only then does Pop leave her side and paddle out to take up his watch in the middle of the pond.

Just before I leave my perch in the tree, a large V of over fifty loudly honking Canada geese comes by flying high in a northerly direction. Peep and Pop remain silent and pay no visible attention to them. Those birds who did not breed should now or soon be migrating to their safe tundra habitats in Ontario for their molt, when they lose their ability to fly for a month. The breeding birds will also lose their flight feathers, but they can't migrate to the molting ground because they must stay to rear their young.

May 31, 2000.

The eggs must now be close to hatching, if any are still alive. I'm hopeful that the one egg that was not rolled out might hatch, and as I examine the eggs this morning, I'm cheered to notice that one has a pip in it. I hold this egg up against my ear, and . . . I hear it peeping! (The other two are silent.)

In midafternoon Peep is shoring up the nest, still pulling nest material around and under her. Pop is standing close beside her. There is no visible progress from the egg, but the still-imprisoned gosling's peeps now sound almost frantic. They remind me of Peep's heartrending cries when, as a small gosling, she was temporarily left alone and wanted (needed) company. The imprisoned gosling could have been trying to communicate either with its mother or its siblings. Precocial birds that are led away from the nest by the mother (or parents, in the case of geese) within hours after the eggs hatch communicate with each other to synchronize hatching. It is costly for a chick to hatch late (or to be silent?), because unhatched eggs are left behind.

June 1, 2000.

I rush down to the bog by 5:30 a.m. with a camera to record anticipated events. Pop is by the nest. Peep has her head held up on full alert as I come through the woods to the edge of the bog. But as soon as I slog through the bog to the nest, she rests her head on her back and closes her eyes; she has recognized me. All is well. She goes back to sleep. I reach under her and pull out one egg after another. Pop preens directly in front of me, slowly and distractedly. Two of the eggs that I handle now definitely stink—they contaminate my hand with their stench. Peep has made no attempt to roll out silent eggs. This time I don't hear peeping as soon as I hold the (third) egg up to my ear. Instead, I hear a rhythmic scratching-scraping noise. So—the gosling has stopped calling and is getting to work, rasping the thick eggshell with the egg tooth at the tip of its bill. I suspect progressively more oxygen is entering the egg now that a chip of shell is off and air has direct access, rather than being limited by diffusion through the pores of the eggshell. The increased oxygen should permit more vigorous exercise, and the exercise of trying to escape from the seemingly rock-solid eggshell may be strengthening the gosling for the rigors just ahead, and so I offer no help.

By afternoon the gosling has made approximately an inch-square hole. The imprisoned gosling now peeps loudly when I pick the egg up. And the peeping continues even after I put the egg back under the goose. Pop comes close and extends his neck down to the nest, as though listening. Peep makes a series of rapid little chuckling grunts. She still constantly pulls vegetation around herself, and tucks it in to her side. I again offer her handfuls of dry sedge that I pluck and she accepts and uses all of them for her nest. She rejects wet, soggy sedge and briefly takes green sedge from my fingers before dropping it. The gosling keeps working. Finally, by the evening, it has broken free: When I reach under Peep, I find an empty eggshell, and a freshly damp gosling! It has not yet had time to dry off. It's time to pop the champagne.

It is already getting dark when I have to leave. But I expect to see the fluffed-out gosling at dawn.

June 2, 2000.

It is barely light, and as I'm preparing to go to the bog. Peep and Pop arrive at the house. I instantly know the bad news: The gosling is dead. They would not leave it alone for even one minute, so long as it was healthy. I rush down the hill to the pond, and they fly down as well.

When I get there, Peep is sitting on the nest and Pop is nearby. I wade out and find one egg floating in the water. I hear no peeps. I search under her and find only one stinking egg. Then I see the gosling—floating dead in the water about three feet from the nest alongside a clump of sedge. Retrieving it at once, I can't see a mark on it. But a gosling doesn't drown by itself.

I'm affected. I've been seduced by the geese. Some biologists are apologetic for forming attachments, because they feel that it makes them less objective scientists. True enough, by concentrating my attention on the geese in this beaver bog I have looked less closely at the beavers, the red-winged blackbirds, the goldfinches, the rose-breasted grosbeaks . . . My focus on the geese has been like a compass that has directed and held my attention. However, I don't feel that this has compromised my objectivity. Quite the contrary, it has prevented me from being captured in fixed mental frameworks that might blind me to what the geese are, as opposed to what they ought to be. I follow *them.* I did not and do not know where they will lead me. The problem of compromised objectivity comes less from being enamored of a beautiful animal than from being too infatuated with a beautiful theory. It arises through being led by an inner light rather than by external, empirical reality. Expectations, hypotheses, and theories are tools that, like the lens of a telescope, direct our view. They can distort and restrict, as well as enhance our view.

After provoking Peep to rise, I put the dead chick under her in the nest. She reacts indifferently to it, shoving it out of the way as though it were a clump of damp moss. I then lift the smelly egg out of the water and put it onto the edge of the nest. She immediately reaches over, puts her bill on the farther side of it, and rolls it back into the nest and under her. Her actions seem pointless, yet they can be explained by evolutionary theory. The goose's nesting is finished for this year. But she cannot yet know this for sure. She is still acting out a behavioral script in response to essential, or sign stimuli.

My curiosity about what the geese saw, felt, and might do or what had happened had barely begun. To attend to first things first, I dissected the dead gosling. Although I found no teeth marks, there were hematomas on the head and neck proving that it had experienced violent treatment while still alive. It was at least the third time that the nest had been attacked, and again it was not likely motivated by a predator seeking food. It was impossible to know who the attacker had been, but there was still opportunity to ask the geese questions. For example, how long would they continue to incubate two rotten eggs?

Not long, it turned out. Peep and Pop came to the house three more times that day. And the next dawn, already at 5:45 A.M., I heard heavy wing-beats below the bedroom window. Pop, for the first time since the eggs were laid, sat on his old sleeping perch, the big rock in our yard. Two hours later he was still on the rock, and Peep was lackadaisically wandering all over the yard, nibbling on tulips, daffodils, a bucket, a blade of straw, the bench. She stopped to stretch. She made her comfort sounds. I rushed down and plunged into the cold water of the bog once more, just to be absolutely sure that the two eggs were really still there, and they were. So Peep would *not* sit on rotten eggs even for one extra day. I thought that odd, because she had indiscriminately incubated the dead, rotten eggs for days, not to mention a rock, a potato, and a kiwi fruit. Had she exactly counted the twenty-eight days normally needed to incubate the eggs? I doubted it. The gosling (because it was alone?) had hatched three days late.

As earlier in the spring, I again scattered grain on the ground for Peep and Pop. However, Peep ignored the grain, and as I wandered around our yard, she followed me. She now ignored Pop, as though her attachment to him had evaporated. Peep had similarly ignored Pop on the two previous occasions that she had (almost) abandoned the nest. Her association with Pop thus seemed like a response directly tied to nesting. Yet the two other geese had spent the entire spring in the neighboring pond, and despite having made *no* nesting attempt, they had maintained an inseparable pair bond.

Peep and Pop flew up to the house once more the following morning. This time they stayed the whole day without leaving even once. Peep again followed me, and also Rachel. She again looked up at me, and did so often. I looked back at her, and one time, when I tested her, we maintained eye contact for minutes. That eye contact seemed strange, almost eerie. I didn't know why she was doing it.

Peep ignored Pop as she continued to follow me around. Not once during the day did Pop make the "Let's go" signal that he always made when she had been off the nest too long, making me wonder if he "knew" when the nest needed tending and when she needed to get back to it.

With nothing in the bog to hold them, I expected that both she and Pop would now stay with us up at the house "for good." I was already wondering how to keep them comfortable through the coming winter. However, when the sun approached the horizon, Pop stood tall, faced northwest, and made the "Let's go" signal. This time she responded with head wagging. She then pushed off, and Pop flew right behind her as both, honking loudly, left for destinations that would forever remain unknown to me.

The nesting season was over, and the goose hunting season and winter would follow. Peep was out of my charge, having vanished into the unknown as she had done two years earlier. She had, however, again left a precious gift. Her gift to me was the unexpected link that now connected me to her species and her world.

5. SPRING AGAIN

It's the twentieth of March, which marks the vernal equinox and the official first day of spring (2001). As if on cue I heard the first red-winged blackbirds yodeling in the still-snowed-in bog. It snowed hard and continually over the next three days and nights. Despite the storm a few male redwings returned daily to the bog, staying only a few hours before leaving to find their food elsewhere. Still more snowstorms blew in on the twenty-seventh and thirtieth of March, but on April 4, I finally heard the most welcome and long-awaited sounds: the calls of migrating geese. I looked up, and against a bright robin's-egg-blue sky I saw snow geese. Thousands. One formation after another of these immaculate white birds with black wing-tips passed over, and all were heading south following the wind. I had expected to see them going in the opposite direction. Had they already been moving north toward the tundra above the Arctic Circle, found no open ground at customary stopovers, and thus been forced to backtrack because of the late snow?

A wind change brought thunder, lightning, and pounding rain on the night of the seventh of April. With continuing swift wind now blowing from the south, even more rain was predicted. The snow yielded, and in the beaver bog, open water collected in front of the upper dam. Here and there the beavers were patting fresh mud up against the dam, which still looked like a wall of snow. They could not fix the dam that way. No wonder they never managed to fix last year's underground leak.

Beaver tracks led from the lodge to the dam, and there was a well-worn path through the snow to a just-felled white birch tree whose many limbs had been clipped off and dragged back to the lodge over and through the snow. These beavers did not have a sufficient underwater storage of twigs under the shallow water to last them all winter.

By the next day, the sun is shining and hundreds of honkers, the Canada geese, are gathering in the fields at the end of Shelburne Pond. Thousands more congregate on other fields in the valley along the Winooski River and near Mallets Bay of Lake Champlain. Many geese are paired, and these pairs are apparent along the peripheries of the great crowds. By afternoon we hear the first honking of returning geese down in the bog.

"Rachel, Eliot, did you hear that?" I holler. "The geese are back!"

We run down the driveway, and there in the middle of the main lower pond where water is flooding onto the ice and soaking the snow, stands a pair. Could it be Peep and Pop?

"*Peep*—hello Peep! Hey Pop!" I yell hopefully, even though we have seen no sign of them for ten months and several days, and have not really expected to see them again.

One of the two geese stands at attention. I keep calling. The goose starts walking through the slushy snow—in our direction! It stops. I call again. It looks—scans—walks forward again, and stops.

I run up to the house to get my snowshoes so I can walk out onto the dam. (I also remember to bring cracked corn.) Without the snowshoes I'd sink over my hips through the snow and into icy bog water. By the time I get back and have put on snowshoes, they are walking on the pond toward the upper beaver bog where Peep and Pop had nested. When they encounter the dense willows and cattails, they fly over them, landing next to the old nest site in the sedges by the beaver lodge. I walk onto the snow-covered dam and keep talking: "Peep—Pop—so good to see you." "It's all right—remember me?"

One of them looks toward me as I get closer, then reaches down with its head to rub it across the chest in a movement that is likely a nervous displacement activity. As I pour grain onto the snow, the goose's head goes up in a small jerk—definitely a sign of recognition. It's the male, and I can tell from his face that it *is* Pop: his unique hatchet-shaped white face patch is unmistakable. I'm seeing a friend after a long absence, and our mutual recognition makes me feel glad. Geese are presumed to mate for life, so he should be with Peep; unfortunately Peep's face is less singular, at least to my eyes. Her white face bib has rounded contours, and like a number of other geese, I cannot identify her except at very close range.

Slowly Pop walks toward me and stops within about twenty feet. She joins him but seems shy, though not nearly as shy as the geese at other ponds. I can't get close enough to her to see her eye scar and the small nicks in her face pattern that would identify her as Peep. As I move closer to her, she becomes nervous and walks away. Peep would not normally have done that, I think, but who else could it be?

I return to visit the bog in the evening as it is getting dark early owing to a thick overcast. Robins sing loudly and melodiously all around. The first of two woodcocks, one on each side of the bog, rises off melt-spots of bare ground and launches into the air, spiraling out of sight into the sky. Its syncopations are barely perceptible over the raucous robins' *Nachtmusik*. But in several minutes the robins stop their compositions and make loud sharp notes that signal the end of the day. Pop and his mate have left, and I now see only their fresh footprints in the mud the beavers have recently plastered onto the dam.

Just as I'm about to leave as well, two geese come flying in from the west and alight on the ice of the main pond. I call but they do not come to me. The woodcocks continue to spiral through the night sky over the bog unleashing their unearthly twittering. Then, they descend on fluttering wings onto some exposed patch of earth to make soft nasal *peents*.

At dawn, when I return, Pop and his mate are in the melt area below the breached dam. He stands up to his belly in the swiftly flowing icy water. She is feeding with her bill under the water and vibrating it like an electric mixer to agitate the water into wavelets around her head. She lifts her head periodically and water droplets glisten on the back of her head. His head remains high and dry as he scans all around and makes a soft purring sound. At intervals he also makes a deep throaty rumble like a kitten's. She is silent. In the past, Peep had not been so uncommunicative. Again he purrs and she, next to him, rises up out of the water to violently shake herself and then settles back down to preen. After a while she tucks her head behind her wing. I see the white lid of her closed eye—but every once in a while she opens an eye, at least the one toward me. He continues to purr. After a few minutes she takes sips of water, and he then does the same. She preens again for a few seconds, drinks, and "sleeps" again while he continues to stand tall, purring and gazing all around.

April 12, 2001.

I stand at the edge of the bog in a drizzle under a dark sky. It feels colder than the 40°F indicated by the thermometer. The female goose is asleep on the beaver dam with Pop alert beside her. He looks toward me and starts to approach, walking in his slow, deliberate waddle that looks dignified and reserved. His mate awakens and follows him. I strew some grain, but back up because they seem hesitant to approach. Does he now rush toward the grain? No—he stands on the dam about ten feet from the grain and waits for her to feed first, even though he initiated the approach. He could have fed immediately, but he held back until she came up out of the deepwater channel in front of the dam. When she finally approaches the grain, she feeds eagerly and he stands watch. He does not look back toward me. Instead, most of his attention is focused toward the woods on the other side of the bog. She feeds uninterruptedly for six minutes without even once looking up. The very second she lifts her head, he lowers his and walks to the food to finally take his share. She scans, and then, after he finishes, she feeds again for about a minute, then walks to the edge of the dam, slides down into the water, and "dips" for food under the water. Pop stands like a statue near me in the rain. I bask in the glow of our mutual tolerance, if not friendship.

Silver water droplets accumulate on his back, coalesce into ever-bigger droplets, and then roll off like round opalescent pearls. Five tree swallows skim the frozen pond back and forth and all around without pause. What could they possibly be finding? Occasionally they dip down onto the softening ice! And after twenty minutes of ceaseless erratic flight, they all leave together on their migration to their breeding grounds.

The next morning I decided to see if Pop would keep watch while his mate napped. I got out my always-present pen and piece of paper, and wrote down every minute, on the minute, what each was doing, for 70 minutes. Throughout this whole time Pop had one leg drawn up into his belly feathers facing her about four feet away and his back to me. He never shifted his position. He scanned in one direction, with his head held immobile. He'd turn his head after a few minutes and again hold it steady to look in another direction. She slept for 57 of the 70 minutes, preened for 6, and watched for the 5 minutes that he preened. She slept so much that Pop did not get to sleep at all.

Later on when he finally tucked his head onto his back, he still kept his eyes open—at least the eye I saw. Once in a while he closed it, but only for about a second. She, in contrast, only occasionally opened her eyes for a second or two when she was in the same sleeping posture of head tucked into back feathers. She did not act like a wild goose, but I still was not convinced it was really Peep because I had not seen her scar. My gut reaction was that her character was "different."

Watching sleeping geese is not demanding or particularly exciting. I quit after a little over an hour when I became distracted by a couple of song sparrows. One was perched on a snow-bent white birch tree hanging out over the bog. He sat in one position, one spot, and he sang his song, repeatedly without fail 8 to 10 times per minute (I timed him), for at least an hour. Throughout this time a second song sparrow in an arrowwood bush faced him from about fifty feet away also singing uninterruptedly, but only five song-sequences per minute. The first bird was by far the louder, and he had a long, lowered note at the end of each song sequence, while the second always added a high, sharp trill to the end of his refrain. Only the first, the one on the bent birch, showed impressive improvisation from one song to the next.

The two sparrows stayed near but separate from each other throughout their song duel. Eventually they both stopped singing, met on the ground between them, and made high, metallic chirps that sounded like someone lightly hitting a piece of glass with a spoon—not the louder *tit*-call these sparrows usually make. After a while the tinkly metallic chirps stopped, and while still close to each other on the ground, they both started singing again, but very, very softly. At the same time, they lifted and vibrated their wings just before singing, and one flew to within a foot of the other, who then moved off a few feet. After about 15 minutes of interactions on or close to the ground, the first hopped higher and higher into the brush. At the same time he gradually became louder in his song, until he was back on the same spot on "his" birch, where he'd sung for 43 minutes previously. He then sounded off as loudly and vigorously as before. The second bird became silent.

Pop and his mate stayed on the beaver dam at their (upper) pond all morning. They were quiet, but near 1 P.M. a second pair of geese arrived and landed in the main (lower) pond, and all four geese honked loudly. Pop was especially noisy, and he gesticulated to his mate by bobbing his whole neck and head till he was almost horizontal in front of her. The strangers both took to the air and flew over toward Pop and his mate, who rose up and chased them back.

Although the other pair returned to the lower pond, they were still not far enough away for Pop. Again he loudly gesticulated, honking and jerking his whole neck and head toward his mate. Pop and his mate again took off, and Pop pursued one of the newcomers, who flew only a short distance, landed, and was joined by his mate and the two walked away, one next to the other. Pop and mate took to the air a third time, and this time evicted them. The intruders left the pond entirely, flying over the lower dam and then down the valley.

Pop and his mate then landed in the now-open water next to the old beaver lodge, and both started to bathe, which I had not yet seen them do this year. With vigorous abandon they ducked down and threw water onto their backs, ladling by a quick jerk of the crook of the neck. Then Pop's mate walked onto the lodge itself, perched on the ice that was nearest to it, and preened.

On April 14, as I was again watching Pop and his mate and admiring their strong pair-bond, I heard loud goose-honks from afar and rapidly coming closer. These calls, down in the valley to the west, were from a goose who knew this place (and its inhabitants?) and was announcing its arrival in no uncertain terms. As the noise got closer, Pop stood tall and attentive, but instead of sounding the challenge that he usually gives to such pronouncement of visitors, he inexplicably remained silent. Usually he mounts his vocal challenge as soon after the noisy interlopers can be heard from a distance. Something was different. What was different now? Then I saw one goose flying in, all alone!

This goose did not stop at the large target, the main pond where practically all incoming geese land, but instead came up to the small upper pond, near *them.* Pop did head-dipping, and both he and his mate flew over to the newcomer. But unlike yesterday it was not he that attacked the stranger. It was *she* who did so, and vigorously, and all alone, and several times in succession. He tagged along, but when he got close to the newcomer, he did not attack. He instead performed a noisy greeting ceremony involving vigorous head-bobbing close to the ground and head-shaking. The display looked much like those head movements he had made to his mate.

After several attacks by Pop's mate, and greetings by him, all three geese preened vigorously. Then Pop's mate took off after the newcomer anew. He again tagged along as she again attacked, but confusion ensued as he joined in, and I lost track in the continuing melee; I *thought* I saw Pop leaving his mate to join up with the stranger. Could this be?

Minutes later I watched another attack, with Pop following his mate, and the new bird, a female, was finally driven off. Uncharacteristically for a beaten goose, however, this one came right back and then crossed the pond to land near me! I was baffled, because this bird was absurdly tame. Why did she come to me? She blinked with her left eye and . . . her face looked like Peep's. Could *this* newcomer be the "real" Peep? My mind rebeled, teetering on the edge of the possibility.

Consistent with my new notion, the newcomer perched alone on the ice close to where Peep and Pop had nested last year. Pop and his (now new?) mate were on the other side of the pond. After about ten minutes Pop's mate again resumed her attack on the single female while Pop again did his vocal and vigorous neck-gesticulation greeting display to her. Pop then calmed down and stayed with her, but his mate busted in for another attack. During still another melee that ensued, Pop grabbed feathers on the newcomer's back and appeared to be repeatedly trying to jump onto her in a forced mating. Afterward Pop rejoined his mate, who again attacked the brazenly persistent intruding female who just would not leave and who was determined to stay despite the sustained abuse she was getting.

Attempted forced copulation.

In the afternoon, Pop and his mate were again beside me on the dam. The other female *still* had not left. Pop and his mate occasionally called to her and vice versa. She stood tall and watched us from about 100 yards away. After a long time, maybe twenty minutes, I saw Pop give the "go" sign to his mate and they both took off directly toward the lone goose. But this time his mate veered off and flew in the *opposite* direction, while Pop continued undiverted. I had never seen a gander leave his goose's side like that.

Upon reaching the newcomer female, Pop again gesticulated in a vigorous greeting display and soon *they* looked like a loving couple. After ten minutes they were still side by side and they started an animated bathing routine, dipping down, throwing water over their backs, and merrily engaging in a mutual bath. It's not a thing for a gander to do with a stranger, while his wife is away.

After the bath *à deux*, Pop walked onto a small rise and as if suddenly remembering, looked around, possibly for his "real" mate, who had just called. She was still out of view, down at the edge of the lower pond, but I soon saw her start walking through the cattails up toward them. When she finally got near, the *stranger* attacked her and then went right back to Pop's side. But not for long. Pop made one more cursory mating attempt, and then joined up again with his mate. (I would later "normally" only see mating just before or while the eggs were being laid.)

By now it was clear: Pop was not going to chase this lone female away. But his mate did not tolerate her. And vice versa. Pop's present mate was by far the more aggressive of the two females, but the newcomer was incredibly persistent. She was getting mixed signals and she was not going to turn tail easily despite the impressive onslaught that she faced.

When I returned in the evening, the newcomer was alone at the easternmost dam, at the bog periphery. On a hunch, I approached her. She paid me absolutely no attention. Strange! At the very least, wild geese are alert and check you out. She dipped her head down under the water, tail up, searching for food. Pushing my luck and acting on intuition, I said: "Peep!—Come here!" This time there were no other geese around to distract her attention. We were finally alone together. She stopped for a second, looked, and then nonchalantly walked toward me. I hunched down, grabbed a bit of weed, and called again. She came closer and stopped. It was now my turn, and I walked up to her, and there, within several feet of me she bathed, dipped, and splashed, and then rose up, and performed the "feel good" wing-beating ceremony that the geese always give as though in satisfaction. Then she walked unhesitatingly all the way up to me, examined the weed in my hand, but refused it. I reached down and stroked her back with my hand. She acted totally unconcerned. I again stroked her back. No reaction whatsoever. Not even a twitch. I now saw the scar on her eyelid. I saw the identifying nicks in the pattern of her white face patch. Now I knew: Pop had bonded with a *new* mate, and the "stranger" who had arrived a week later, was Peep.

I feel enlivened by Peep's presence. I've been rooting for her for a long time and now I see that she has lost Pop, her old mate, and I'm quite sure she'll also soon lose her pond, her home. She can't withstand the continuing attacks of Pop's mate, at least not without a strong ally, a mate of her own.

6. MATE SWITCH

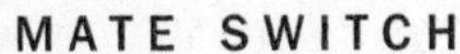

It was no accident that Pop's new mate was tame. She turned out to be the original resident female of the bog, who last year had nested on the muskrat house in the lower pond. I now christened her "Jane." I had not paid much attention to her and her gander last year, since I had concentrated most of my attention on Pop and Peep. Luckily, I had photographed her on the nest, and when I now examined my accumulating stash of mug shots, I was surprised and pleased that despite the intervening annual feather molt, the same specific nicks and notches, and even the same unique black freckle, were still found at the edge of the right side of Jane's white face patch. Apparently she had lost her mate, and so Pop had either bonded with her in the flock and then returned to the bog with her, or they had both returned independently and then bonded. Geese pairs normally stay together, and I suspect Peep's dissolution from Pop could have been related to a weakness of the pair-bond owing to her imprinting on humans. Her bonding to place should not have been affected. Being divorced, she still came home, and she needed to have a new mate find her quickly in order to breed.

Are geese monogamous? I think the answer is yes, and no, depending on circumstances and on definitions. Geese and other waterfowl are highly philopatric; they return to nest year after year at or near the same place where they have nested before. If both members of a pair return to the same "house," then that behavior alone would make them functionally monogamous. Had Peep returned to the bog several days earlier, and Jane later, then Peep and Pop might still be a pair. "Home" is not the only mutual attraction in geese—timing is "everything."

Pop still recognized Peep and showed attraction to her. He was in a quandary. But Jane, being older and an experienced nester, and probably advantaged from her position of seniority, ultimately decided the issue.

April 15, 2001.

A huge commotion has started during the night down in the bog. It continues at dawn as I watch the drama unfold. The excitement has been instigated mainly by one lone gander who has been calling nearly incessantly in the larger main pond. His white bib is broad like that of Jane's mate from last year, and it is without the sharp angular pattern of Pop's. This excited gander stays far from me at the opposite end of the large pond. There is a female near him but she does not have the calm demeanor of a married goose, nor do the two act like an established couple. Pop and Jane were absent when I arrived and it seems odd for this particular male to be so excited since there are no geese nearby to pose a territorial challenge.

Pop and Jane don't return till almost 9 A.M., flying in from the west. Right after they land on the ice, Pop runs to Jane and performs the bowing greeting ceremony. Then, with hardly a pause, he flies up to chase the new male and attacks him viciously. He tries to nip him even in the air and slashes at him with his wings. Uncharacteristically, Jane stays by the lodge. I have seldom seen Pop so aggressive and agitated, nor Jane so hesitant. Pop chases the male well away from the pond, comes back, lands on the ice by the lodge, and quickly runs across the ice back to Jane, to once again perform the noisy bowing greeting ceremony. She answers his vigorous greeting, and I suspect their relationship is cemented. They are now indeed a married couple.

My attention shifts again to the female with the noisy gander. She has started to follow the new male that Pop so viciously attacked and chased off. As I suspect, it's *Peep!* Pop seems overly aggressive to the newly arrived male, and on the basis of his white face bib, I suspect that it could be Jane's old mate (I call him Jack). If so, perhaps Pop should recognize him as an especially great threat. After all, Jack may still want to claim the female that he, Pop, has now "stolen" from him. But since Peep is now showing interest in Jack, perhaps that threat will be deflected. Could the two former pairs end up switching partners?

Next, Pop takes off and attacks Peep, then flies back to Jane, and both go through their noisy greeting ceremony again. He is reassuring Jane that he now belongs solely to her, not to his former wife. Then Jane, *on foot*, comes toward Peep, who has come near me. She runs and he follows. They go right past me as if I am not even there, and she attacks Peep, who flies off. Now *both* Jane and Pop pursue Peep until they have driven her down below the main pond and out through a gap in the trees to one of the small adjacent ponds to which Jack had escaped earlier from Pop. Then, instead of returning all the way back to the upper pond, Pop and Jane stop at the old lodge where she naps and he stands with head held high to keep watch.

By noon Jack and Peep get pushed by Pop and Jane to the bog's periphery, up against the willow bushes in the east. There they stay near each other, and now having a common enemy, they almost seem like a couple. Pop and Jane remain over 100 yards away. Pop is now the only noisy, aggressive one, flicking his head as though trying to induce Jane to join him in another chase. She refuses. He pecks her on the head for emphasis or to get her attention. She still doesn't join. What's going on? Is Jane refusing to attack Jack and Peep because Jack is her old mate? Pop seems to be trying to herd Jane, his new love, away from Jack. Eventually Pop takes off after Jack (and Peep). Jane follows and attacks Peep, who runs into the bushes and hides while Jack flies up and Pop pursues him aggressively. But Jack only flies to the new nearby lodge. Pop lands near him, and Jane, who has followed, does so, too. Pop now (temporarily) ignores Jack. He runs to Jane—being very attentive to her by performing an intense greeting ceremony. After this interaction Jane settles down for a nap and Pop stands and watches, with Jack only about 100 feet off.

Peep, meanwhile, is nowhere in sight. But I notice Jack continually looking toward the bushes she was chased into by Jane. After a while Peep calls weakly from her hiding place there, and Jack answers excitedly and starts walking in her direction. Now Jack is becoming the noisy one and he is totally oriented in the direction from which Peep has called. She stays silent and hidden. But Jack keeps walking toward her. Eventually he reaches the smallest, easternmost pond of the bog where Peep has found refuge.

I find them both still there an hour later, close together. She is dipping and bottom-feeding. He has his head up, alert and watching. They are beginning to look and act like a pair. But this little peripheral pond, a puddle really, will not be suitable for nesting. It is a hopelessly marginal habitat. I hope they'll not be driven away from the bog, but I fear we might lose Peep and her new mate.

Jack and Peep are still together by evening, and they are still at the same little peripheral pool. As I watch them, I feel glad that Peep has apparently found a mate, even if it is Jane's old gander. Jack already follows her as she feeds, and then I see him take off, and she rises into the air and follows him! They fly directly over Pop and Jane at the upper pond, Peep's old nest site, and then they continue down over the lower pond, and finally pass on into the valley. Inexplicably, they then turn around and fly back once more. Jack lands in the middle of the big, lower pond, his former home, but Peep flies on up to *her* old home, where she joins Pop and Jane. But this time Pop makes his greeting ceremonies only to Jane, who then attacks Peep. Clearly it's all over between Peep and Pop. Peep then flies back down to the lower pond to rejoin Jack. Pop (with Jane following) starts once more on the offensive toward Jack, but seems to lose his nerve when he gets close and lands near Jane instead. He runs to her and vigorously performs the greeting ceremonies. Peep is separated from Jack and calls. Jack stands tall, faces in her direction, answers, and then does the head-flicking, pretakeoff motions. He rises and flies to rejoin Peep and they leave the bog, flying together side by side.

Peep and Jack returned once more the next day. Pop and Jane again routed them into the woods onto the same tiny peripheral pool. Again, Peep and Jack did not stay long, and both left together as they had yesterday. Pop and Jane now owned the entire bog.

I checked (but not exhaustively) at neighboring beaver ponds throughout the spring and summer hoping to glimpse our Peep and her new mate, Jack. I saw numerous pairs of geese, but not them.

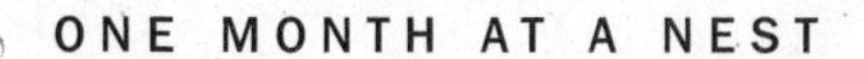

7. ONE MONTH AT A NEST

The female picks the nesting spot in the world of geese, as in the worlds of most other bird species. And female geese and ducks often come back year after year to or near the same spot to build their nests. I therefore expected that Jane would soon start to nest near or on her old nest site, the muskrat lodge in the lower pond. I wanted to watch the family from up close without having to wade through mud and leach-infested water to that rat lodge, and I therefore tried to entice her to choose an even better alternative site, one a little closer to shore but near her old nest site. I made a solid nesting platform of water-soaked logs, amply filled it out with hay, and then topped that off with dry sedge. It seemed like an ideal little island and a much better nest site than the soggy decaying muskrat lodge.

The pair noticed it. By the next day they had pulled all the sedges off into the water, and then they rejected my generous offering. (But it was chosen by a pair of newcomers to the pond, the Sedge pair, the next year.) Jane again chose the same, and to my mind vastly inferior, now-flattened muskrat lodge where she had nested the year before.

The first time I saw Jane on the flattened muskrat lodge (until then it had been used as a perch by mallards and a great blue heron), she had her head up and her bill slightly open—I presumed (correctly) that she was probably laying an egg. She had started laying her clutch with very little or no nest-building preliminaries, and after laying six eggs started her month-long incubation.

April 25, 2001.

Jane is on the nest at 6:10 A.M. when I arrive, and when she sees me coming she gets up and covers her eggs with the surrounding debris of decaying cattail leaves. Pop then comes to me and she gets off the nest and follows. Their backs are tinged with hoarfrost. A lone mallard drake swims to the cattail point and comes within six feet of their nest. Pop and Jane continue feeding next to me. But after six minutes Jane abruptly lifts off and flies directly at the duck, who leaves in a hurry. Pop then approaches Jane and after a noisy greeting-ceremony she paddles back out toward the nest.

A low mist drifts over the water by the nest. The redwings are numerous and raucous in the cattail portion of the bog, and next to me three ruby-crowned kinglets forage silently in the arrowwood bushes. One hops near my face and the vermillion crest on the top of its head shoots up like a flame. The snipe's monotonous accipitrine *tucka—tucka—tucka* call tolls from the border between the cattails and the sedges. Pausing briefly in its exclamations, the snipe flies up, sets its wings, and immediately drops back down into the sedges. Then it is silent.

April 26, 2001.

Touched by the first rays of the morning sun, the frosted sedge hummocks morph from white to rich brown. Flickers call and a ruby-crowned kinglet again forages near me in the arrowwoods. A by-now-familiar male redwing (one with a stub tail) is back. As on previous days, Stubby stays between the elm to my right, the cattails by the goose nest, and the bushes to my left. Curiously, I never see him chase any of his neighbors. Isn't he supposed to "defend his territory"? A pair of mallards perch on the old beaver lodge, and then a single male on whistling wings flies in and slides onto the water with a resounding *whishssss* at the east side of the pond. The pair flies up, and the single male joins them, as all three—two drakes and one hen—make several passes around the pond before leaving together. The lone snipe continues its syncopated call from the marshy ground where the beaver-cropped willows and sedges intermingle.

Pop and Jane paddle around near the nest and dip for underwater food. Then, at 6:53 A.M., Pop slowly and quietly glides over to me. She follows. But after about three minutes they gaggle loudly, and she hurriedly returns to the nest. She walks up onto it and while standing, preens. He perches next to her at the nest edge and also preens. At 7:02 she settles down upon the nest, gently fussing with her bill all around her body as a piece of down adheres to her bill. Pop floats in the water close beside her. Jane stays on the nest all day. Incubation has probably begun, although the goose could be sitting on the eggs without yet incubating them (without heating her feet and making skin contact with the eggs).

At around 5 P.M., I wade out to check the nest. At first Pop stays at a distance and lowers his head, as if to hide. As I approach the nest, he swims up to me but remains silent, and when I get to within about six feet of the nest, he crowds up to me and starts to hiss. Jane sits tight. I break off a dry cattail stalk and try to nudge it under her. After a little poking she stands up—I count five eggs—already a typically full clutch for a goose of four years or older (but she lays one more egg the next day). Then I withdraw. I have barely retreated ten feet when Pop stands tall and powerfully beats his wings in the goose's gesture of power and satisfaction. His wing-beats sound like a powerful drumroll.

I return to the pond in the evening as a thin crescent moon shines high in the sky to the west. The spring peepers start their ear-throbbing chorus, and throughout the night, whenever I crack the window, I can hear the chorusing of these diminutive frogs.

April 27, 2001.

This morning the peepers are silent. A southwest breeze ruffles the water and Jane appears to sleep on the nest with her bill tucked into her back feathers. However, her eyes are wide open. Pop paddles over to me from mid-pond to feed on a handful of corn. Jane raises her head, rises to her feet, fusses for a minute or two covering her eggs with vegetation from the nest edge, and then lifts off and flies about 20 feet out into open water. From there she paddles the remaining 100 feet to Pop and me. As soon as she arrives and begins to feed, he lifts his head and watches and makes soft, low, gaggling noises interspersed with little purrs that can only be heard by standing alongside. His soft chatter probably reassures her that she can feed without having to interrupt and scan for danger. Done feeding, she leaves abruptly and swims fast toward the nest.

Her feeding and swimming tempos are no longer slow and languid as before. They have both become jerky since incubation began. Nevertheless, she still has time for preening, primping, and bathing, which begin when she stops ten feet from the nest. Pop comes toward her, paddling even faster than she did. She then rushes onto the edge of the nest, preens some more, climbs on, and then preens steadily from 8:05 to 8:15. Finally she settles down to resume incubation and Pop then moves off around the cattail point and into shallow water about fifty feet away.

Much of what I see is by now becoming repetitious and familiar to me. This is something to strive for because without the attainment of familiarity, the significant remains invisible. Immersion in the familiar also brings me a feeling of tranquility and comfort, a sense that all is right with the world.

April 28, 2001.

I awake at 4:30 A.M. when it's still dark, having kept my window partially open hoping to be awakened by the snipe's whinnying during its sky dances over the bog. His is a spring sound that, like the woodcock's, stirs me. Although I have commonly heard snipe display in the sunshine at noon, this one stops his aerial display near dawn when the robins start to sing and when the crows sound off, just before the song sparrows' first chantey.

I'm at the pond at 6:00 A.M., long after this snipe has come to rest. Two beavers are still out and about. One rounds the cattail peninsula and swims past Jane, who is sitting on the nest with her

The view in early May of the bog's larger, lower beaver pond from my usual observation post. The view here is to the southwest. The water line at the farther edge of the pond (to the left of the middle birch tree) marks the top of the beaver dam. The dam is maintained by the beavers from spring through summer, and it is overgrown with *Viburnum* bushes. The beaver's lodge is just out of view to the left of the picture.

A close-up of Peep's head while she incubates on her nest.
The upper photo shows the scar on her left upper eyelid.
In the lower photo, her lid is pulled up over her eye as she sleeps.

Nests of four birds found in the bog. From top left to lower right: willow flycatcher, red-winged blackbird, song sparrow, and catbird.

Top: Peep and Pop during a standoff against a neighboring pair of geese. Pop is the nearer bird in each picture. Note his injured and displaced wing feather, the result of an earlier fight in the continuing interaction.

Bottom right: View in late April of the bog's upper, smaller beaver pond. This is the area that Peep and Pop claimed. In the lower left foreground of the photograph is Peep, almost hidden while incubating her eggs in a nest scrape on an abandoned muskrat lodge. She is surrounded by spirea bushes that have not yet leafed out. Note the sedge hummocks in the water with fresh green shoots emerging through the dried foliage of the year before. The beaver dam of this pond extends the length of the photograph (and beyond in both directions) immediately behind the lodge.

Jane brooding her goslings in her nest on an abandoned muskrat lodge on a cattail point reaching into the lower pond. Note the shell of a just-hatched egg (May 24, 2001).

Left: Jane and her newly hatched goslings (May 24, 2001). A tiger swallowtail butterfly visits the goslings, apparently attracted to their bright yellow color (a sexual attractant for this butterfly species). A moment later the goslings peck at it, then Jane catches and swallows it.

Right: Jane with two of her five goslings. The youngsters have crawled out from under her to romp on the nest.

Harry (standing) and Jane at the moment when their goslings
first venture from the nest on the beaver lodge of the lower pond.
The gosling at right is about to jump into the water (May 15, 2002).

bill tucked into her back feathers. Pop is out in the open water a couple hundred feet to the west. As I call him, he starts to slowly swim toward me and I notice his back feathers whitened with hoarfrost. He stops after coming within about ten feet of me and remains there reticently even after I've dumped out the grain. I back off into the bushes but he still hangs back. Even at 6:03, when the sun hits the tops of the ridges, he continues to remain motionless.

Finally after a few more minutes Pop slowly walks onto shore and daintily feeds for eight minutes, then leaves, not for the nest where Jane sleeps, but back out to open water where he starts bathing vigorously enough for me to hear the splashing. Bathing, preening, more bathing. His day has started. He raises his body up out of the water, and after a few loud wing-beats eases back down, performs a tail shake, a headshake, another tail shake, and then resumes preening. She sleeps on.

Pop and Jane now stay only at this lower pond, where the beaver dam is always in good repair. Neither ventures to the now-unoccupied upper pond where Pop nested with Peep last year. The upper dam is still leaky and that pond is again shrinking dramatically, like last year.

May 2, 2001.

Cool, moisture-laden air has precipitated dew onto the blue-green sedge shoots that are poking up through last year's matted brown leaves. Overnight the poplars have unfurled their tender, pea-green leaves and the red maples on the hillside on the other side of the pond, which have been a uniform gray for months, are flushing out in shades of orange, red, and purple from the swelling and opening of flower buds.

As on most recent dawns, a male common merganser in brilliant white nuptial plumage and a female with a russet head cavort on the beaver pond. The couple seem exuberant in their constant splashing frolic. At 7 A.M. they are done with that and, flying side by side, they leave the pond. A female mallard and her male escort appear a little later. They soon depart as well, and after a half hour I hear geese coming.

A pair appear over the treetops and set their wings to start their descent. By then Pop is already airborne to meet them, and they circle the pond only once before leaving. Pop splashes back down in mid-pond and resumes his vigil as Jane continues sleeping on the nest, which is now fringed with her recently shed down feathers. I call Pop and he drifts over, feeds by me for a couple of minutes, then swims back out onto the pond. He seems to watch me for a half hour as I stand still and watch the bog and then walk along the old beaver dam, where the jewelweed seeds from the flowers pollinated by ruby-throated hummingbirds were liberally strewn about last summer. These seeds are now sprouting pale-green seedlings. I notice a painted turtle next to me, paddling slowly along the muddy bottom. It has probably only recently crawled out of the mud from its winter hibernation.

The grackles have started to build their nests. As with the geese, only the females do this work, but when a female grackle lands on the cattail tips to look around before descending to her nest site, she is often accompanied by at least two males. Her companions pay obvious attention to her and seem to disregard each other.

This morning I see *seven* male grackles aggregate around one female with grass in her bill. They perch on vertical dry cattail around the female, and they look handsome in their brilliant metallic-purple sheened plumage, which they each show off at short intervals by doing their macho display: a squat, a fluff, and a squawk. They wait and watch while she works on the nest, then accompany her as she flies off to fetch more grass. Meanwhile, the male redwings violently chase females out of the cattails and into the woods and back again. All fly at full speed. Other females are ignored.

May 3, 2001.

Although I have no grain with me this morning, Pop starts to approach me as though expecting a treat as soon as he sees me coming down the slope to the pond. He keeps looking up at me, and I at first stay back. Finally I walk all the way down and he then walks up to me, looks at me, waits a bit, and hisses. It is a soft rebuke, yet a persistent one because I have rewarded him in the past and he now has expectations. He has me pegged for a sucker. It feels reminiscent of getting a phone call soliciting for a charitable donation *because* you had made one before.

Just then I hear geese calling from afar. They almost always give notice on any approach. A pair makes a high pass over the bog and Pop shows no interest. But the pair circles and comes again, this time intending to land. Pop lowers his head, honks, takes off, and routs them before they even have a chance to start their descent.

May 4, 2001.

The fresh green of unfurling leaves is all around, and the sky casts a diffuse light through a thin overcast. It's warm, and the water is dimpled here and there by fish and insects touching the surface. Wavelets in V patterns etch the pond surface as muskrats and beavers scull along with just nose and ears showing. Chalky green cattail shoots poke through the tangle of last year's tan, dry leaves. Only a small cohort of spring peepers are still piping their high, thin notes while pickerel frogs give their raspy, snoring calls.

Beavers have clear-cut the hardwoods (red oaks, hop hornbeam, white birches, serviceberry, sugar maple) and the stumps of these trees are sprouting a thicket of brush that is almost ready for another harvest. This thicket at the pond edge is now an ideal habitat for rose-breasted grosbeaks, and one male is already back, singing his languid song. This spring's first northern oriole is singing in the large maples that have outgrown the danger of being felled by beavers, and from the spiraea thicket at the edge of the bog, I hear a yellow warbler and the first chestnut-sided warbler. The familiar thrum of life is returning all around.

Far out on the pond Pop floats with his head held high and his attention directed to me. I'm not yet at the edge of the pond, where usually I feed him, but he paddles across, walks out of the water, stands tall, and looks up at me for minutes. I continue my reverie. Finally he turns around, reenters the water, and floats back out into the pond. Jane, meanwhile, sleeps on the nest, but lightly; sometimes she opens her eyes even as her bill is tucked into her back feathers.

The nest, on a cattail promontory, is at a crossroads. This morning, during my two-hour watch, a muskrat sculled across the pond and crawled out onto the goose nest mound of mud and decaying cattails. Jane opened her eyes, closed them again, and a second muskrat that sculled by later got no more attention. A beaver fed on cattail roots next to the nest and Jane looked at it in what seemed like mild curiosity. Another beaver came within two feet of the nest, and Pop approached only when it started to crawl out of the water. He then made a splashing lunge at the beaver, who leisurely turned and swam back out. As I expected, Pop then rose up tall out of the water and beat his wings.

May 5, 2001.

A pair of geese flies over the pond and Pop and Jane at first stay nonchalant. The pair lands in the upper pond, where they are not visible, but Jane immediately stands up in the nest and with her bill pulls some of her recently shed down from the nest edge to cover the eggs. Pop swims over to her and says "Let's go" (performs the neck-extension, neck-flexing ceremony). Heeding his summons, she hops off the nest and both start to swim toward me, but then as if remembering the intruders, Jane suddenly flies off toward them at the other pond. Pop follows, and they rout the other pair. In a few seconds they come back and resume feeding, preening, and bathing.

After she returns to the nest and he to the pond, what looked like a slowly rolling log catches my eye. But then a big paw with long claws emerges out of the water and a long, scuted, and tapered tail confirms—a snapping turtle. A head, then two heads at once: Two huge snapping turtles are engaged mid-pond in some sort of a prolonged social interaction.

May 7, 2001.

It's 5:20 a.m. and a bright full moon is still up when I arrive. A beaver is gnawing on the red maple trunk that I have deployed as a support to help me get through the deep mud and out to the goose nest.

Curiously, Jane is not asleep. Her head is up this morning and her eyes stay wide open. I scatter grain and call Pop, but he does not come. Instead, he paddles toward the opposite shore, and his head is held high. Following their gaze I notice a patch of light fur appear and then quickly vanish through the bushes on the far side. Pop stops along the shore where the animal has been and begins to honk. He keeps the honking up for about ten minutes all the while swimming back and forth along the shore and paying no attention to a nearby great blue heron. Apparently satisfied that the danger is past, Pop eventually comes straight across the pond to me. But he only stays for about five minutes, then turns his back to me and again looks across to the other side. She does, too, and then so do I. Now we see a shepherd-sized coyote standing in full view. The coyote remains still and looks all around. Jane drapes her neck down flat over the nest rim and becomes nearly invisible. After about ten minutes the coyote slowly, methodically walks a few feet, stops to look around, continues to weave in-out among the sedges at the pond edge, and stops and looks again. He eventually disappears into the dense arrowwood bushes through which beaver canals lead up onto the wooded hillside.

The beavers did not fix the upper dam, and there has been no fresh beaver sign there lately. No wonder all the beaver activity is now at the lower pond. They have, for this year, abandoned the upper pond, dam, and lodge where they had overwintered last year, and that pond is now drained.

If the geese had chosen any nest site on that pond, as Peep did last spring (a much wetter spring), then their eggs would probably have been eaten by the coyote today, if not much sooner. The upper pond site is not suitable for geese right now, but it could soon be an ideal habitat for even more swamp and song sparrows, willow flycatchers, yellow warblers, and yellowthroat warblers, all of which have colonized other nearby beaver-created swamp.

The grackles' nest-building is done. I see a pair of them mating, and an audience of a half-dozen other grackles appears as if out of nowhere. Following this event, which only takes a second or two, I see a lot of confused chasing. (Other grackle matings attracted no attention.) One female redwing flutters her wings like a baby bird begging for food and while vibrating her wings, she also makes peeping noises. This is a common mating-solicitation display in birds although this time I see no male response. All at once the redwings stop their yodeling songs all over the bog, and as though on signal, they chime in to make repetitive, sad *tiew—tiew—tiew* sounds. It appears like a community response rather than signals from a random collection of independent birds who presumably have no interest in common.

May 11, 2001.

Four beavers are still up at dawn, all four of them gnawing on the red maple log by Jane's nest. I hear soft, mewing utterances through the crunching sounds of teeth on hard wood. Pop passes by them as he swims toward me, and he ignores them as does Jane.

The beavers disperse at first light. Two drift off to the old lodge, which now has piles of shiny, debarked logs stacked onto it; it's the new house for this colony of beavers and will not likely be used by geese this year. It is too steeply conical for a nest platform. Within the hour, a beaver twice brings bundles of green sedge to this lodge. At least I presume it is a beaver. All I see is the bundles of sedge proceeding as though self-propelled on the water surface, and then submerging just before reaching the lodge.

The thousands of sedge tussocks all around the bog are now each issuing dozens of inflorescences. Pop reaches up and strips the flowers of one after another one. There is an infinite supply of these flowers but they are not as conspicuous as the saffron marsh marigolds that now dot the bog, and the even more flashy wild blue iris that will appear in a month.

During this season birdsong reaches a crescendo. A Maryland yellowthroat—the warbler of the swamps and other wetlands—is chiming next to me—*witchety, witchety, witchety*. A catbird—the mockingbird of the bog—scratchily meows and mimics bits and phrases of other birds' songs, patching them together into sometimes jarring but original renditions. Today for the first time in a long while, I hear the muscular drum of the pileated woodpecker.

Beaver bog plants: June 14, 2002.

The sapsuckers and the wild turkey are noisy as always. A kingfisher perches, silent and still on the dead pine standing by the water. It has no "song" but rattles off a coarse chatter just before landing when it flies in. The bird is now poised, ready to dive for a minnow. A commotion of blue jays erupts in woods on the other side, and I see a raven flying through the trees with a blue jay (probably one whose nest has been raided) chasing it.

The grackles of the small colony at the cattail point still travel to and from their nests in *groups* of about four. This morning I saw four fly in from outside the bog with two more joining in, and then six males again perched all around one nest! A few minutes later the six gathered at another nearby nest. Then all six left the bog together as a group. Now I see one, two, or three grackles in the distance outside of the bog, but they join up again before they come back. What's going on? Why do male grackles gather in groups, watch each other's nests and sometimes matings, and come and go in groups?

Pop is, as usual lately, about as far away as it's possible for him to be from the nest and still be on the pond. But as soon as he sees me, he comes over and stays next to me. We both stand still, scanning all around. We both pause to look and listen and we have similar slow, deliberate movements. I like to be in his company, and I don't think it's preposterous to suppose that he likes to be in mine.

May 14, 2001.

As last year, the male redwings seem oblivious of each other, sometimes perching a foot or two apart. But then, inexplicably I saw one male with only small redwing shoulder epaulets (they can be reduced in size by covering them with neighboring black feathers) fly over from the far side of the bog, and this particular individual was immediately chased by a series of other males. Female redwings are still being chased by single males. But most of the time now the females make showy and slow, fluttering flights during which they chatter loudly. Such conspicuously displaying females that don't hide are never chased.

A female redwing with her bill full of nesting material drops down into the cattails, and immediately five males appear as if out of nowhere and a noisy melee ensues and all the males give chase. The chased one dives down and disappears from my view. (A little later I see her or other females with nesting material twice more, and they are *not* chased.) Right after the chase another female perches in full view—the males ignore her. Apparently, not all females are treated alike. Is *individual recognition* a key in their interactions, and if so, on what basis are individuals discriminated against, or tolerated? And what makes it possible for nearly instant coordination of presumably widely dispersed males in different territories to all come together at a moment's notice? Are there intruding females who would destroy nests of competing females?

May 16, 2001.

Pop again comes up to me and then stays beside me this morning. As always we both watch the bog, expectant, open, and relaxed. Anything can happen, and this morning I can't believe my eyes—I see what looks like a snake that is partially erect and weaving back and forth in front of me. But it's no snake. It's a tiny, thin weasel—and I'm looking directly at its white belly. In a second or two the animal disappears.

The swamp sparrows' chatter stops and I hear a chipmunk making high-pitched *chip* calls. Then another and still another chipmunk chimes in, making lower-pitched *chuck* calls. My eyes then catch a flash of brown—mouse? No—I see it again: The weasel is back! Weasel or weasels pop up here and then there through the densely matted sedge. I see something else—a tiny flash of gray mouse fur near where the weasel last disappeared. It's a dead field mouse, *Microtus*, about as large (or small) as the weasel itself. I pull it out into the open, noting that it is still warm and limp. So it was just killed. Nothing looks torn, but blood is dripping from one ear.

I put the mouse on the ground in front of me and Pop. The weasel dashes back and forth, finds the mouse, and pulls it into a hole in the labyrinthine cover of grass and sedges. I reach down and pull it back out by the tail. The weasel lets go of the head end, and disappears. I pull out a shoelace and tie it to the mouse's tail, so I can retrieve the mouse when the weasel tries to run off with it. This could be *fun!* And almost immediately I'm in a tug-of-war with the weasel: Every time it pulls the mouse into its sedge labyrinth, I pull it back out. The weasel persists to about a dozen of these unequal contests, then stands up on its hind legs and looks me in the eye. In a whisk it disappears and reappears on the other side of me and then starts dragging the mouse in the opposite direction—until it reaches the end of the tether. It lets go of the mouse's head, comes forward (toward me), yanks on the shoelace that's attached to the mouse, and goes back to the mouse, grabbing it in the middle and trying again. I fasten the shoelace to the end of a sapling and go up to the house to fetch my camera.

Several sparrows and warblers are scolding the weasel by the time I get back. Again the weasel disappears, but within several minutes it pops up out of the grass—now holding a small meadow vole pup. I make a sudden movement hoping to startle the weasel and make it drop its new prey. It does. The still-warm baby vole is as undamaged as the adult was, and I put it into my pocket. Within a minute the weasel starts searching for its vole(s) and finally ventures to check at the edge of the water, from where Pop has approached and is closely watching. Pop's head rises higher as the weasel comes nearer. To him this mouse-sized animal is of much greater interest than a muskrat or even a forty-pound beaver. It is, of course, a predator on the hunt. Perhaps its movements pique his interest. However, Pop is not afraid. Instead of retreating, he comes wading out of the water and walks up to the weasel, who bounds off. I wonder how he would have behaved if his goslings were nearby.

Jane has been incubating on her nest well within sight of us. She covers her eggs and flies to the opposite, southern end of the pond. Pop immediately honks and leaves me and the weasel, and flies over to her. She preens, and then flies off again to the middle of the pond. He again follows her, honking. Then he takes the lead, swimming fast and directly toward me. Now she follows. When he gets within about twenty feet of me, he stops and raises his head to watch as she comes forward to feed. He had plenty of time to feed while he was with me and the weasel. But he didn't. She hurriedly eats a little grain and some sedge greens, then returns to the nest. She doesn't look around. That's his job.

Dark clouds are billowing past when I return in the late afternoon. The trees' branches rustle and the pond surface is rippling. Jane seems unusually alert and high-strung. Her head is held high. Pop is not in sight, but soon after I call him he comes paddling across the pond toward me, but then he uncharacteristically hesitates. Does he need a signal indicating that I have food? I pour all my grain out—so he can plainly see my gesture. He starts to come forward, then stops again. And then he stays absolutely still, without budging his body, for at least thirty minutes as his head remains high and he scans the shore. He seems more cautious and alert than ever before.

Forty minutes later I finally pull back into the woods to see if it is I who have inhibited him, and not some potential predator. I'm pleased to see that instead of immediately coming closer, he now starts to leave. So I come back down, and he aborts his leave, comes back, and feeds. Apparently he feels *safer* with me than when he is alone. He is, in effect, using me as a guard. But he does now have fears and he is cautious, because he lays himself almost prone onto the water as if trying to hide even while he feeds. He has not done that before.

Pop's unease and heightened alertness make me wonder if the young are about to hatch, so I wade out to the nest to check. He approaches and hisses at me when I come near Jane. Jane, on the other hand, remains passive almost until I succeed in reaching under her. She does not hiss until the moment when my hand touches her belly. I hold one of her six eggs to my ear. There is no peeping in the egg, but it does not pass the sink-in-water test. The time for hatching of the eggs, and then the family taking their leave of the nest, is indeed close. I am now reassured that I can get close to the nest and can look forward to seeing the whole family soon, hopefully to witness one of the most important events of their lives. Pop is sensing or is instinctively aware of coming events, and I suspect that his keeping a low profile acts to reduce potential predators' attention to himself, and hence to the young that will soon be with him. His antipredator behavior is apparently activated in anticipation of the increased responsibility he will shoulder when the young arrive and have to leave their safe island for the dangerous shore.

8. GEESE HATCHING

With Jane's six eggs about to hatch, I come often to be near the nest and I come early because I don't want to miss the event.

The night fog is lifting at 5:30 A.M. in the cool dawn. The sedges and Pop's back are bejeweled with silvery droplets of water. Next to me a robin sings softly as his female keeps crouching down on the horizontal bare fork of a large branch in a white birch tree. She leaves briefly and comes back to the same spot carrying a dry weathered leaf. She places the leaf where she had previously perched in the fork, but the leaf slides off. She flies down after it, catches it in midair, and brings it back and replaces it into the fork. She repeats this exercise in futility again and again. I have caught the very first steps of her trying to build a new nest. (Her other nearby nest with eggs has just been raided, probably by a weasel or a blue jay.)

Suddenly, at 6:12 A.M., I hear a harsh, insistent, oft-repeated alarm call (possibly a redwing's or a grackle's) from the cattail point. Redwings and grackles converge from all around, and within seconds there is a fluttering horde of at least twenty birds at that spot in the cattail thicket. Even Pop, who has parked near me, raises his head and swims over to check on the commotion. I strain but don't see any object of the birds' alarm. The hubbub dies down in about three minutes and the birds disperse. If the commotion had been caused by a mammal, then the birds would have continued to keep track of its movements and to chase it as it left. Instead, the "predator" vanished.

This evening (May 20, 2001) the pond surface is mirrorlike with a spotty rash of insect dimplings. The sky is clear, the air is balmy, and there is no wind. Five beavers meander over from their lodge, enter the cattails, and pull up and noisily munch on the new green blades. The peepers start their shrill chorus and by 8:30 P.M., just as the first fireflies begin to flash, it has reached a crescendo. Bullfrogs *garumph*, green frogs chime in, and a lone gray tree frog trills from the woods near the edge of the pond. Farther off in the hills a coyote begins to howl.

Pop and Jane are watchful. Their heads are up. Jane alternately stands up and sits back down. They may already hear the peeping of the young within the eggs.

After dreaming of geese at night, I'm back at dawn. I hear a turkey tom's *gobble*, *gobble* refrain on the hillside south of the pond, and see a raven with something dangling in its bill fly toward its nest on the cliff. A snapping turtle's head breaks the pond surface for a few seconds, then slowly recedes.

Pop appears from the far end of the pond right after I call him. A mallard drake also flies in and, with his orange legs dangling, circles down and lands about 100 feet from Jane, who is on the nest. Without giving warning, Pop flies up and chases the duck. This is the first time I've seen him act aggressively toward a duck. The world will suddenly become a more dangerous place for the geese when the vulnerable young appear, and Pop seems to anticipate this.

Jane comes off the nest and flies to Pop. This morning she does not come to the grain but forages near the nest in the cattails, bathes, preens and wing-flaps, then hops back onto the eggs and preens some more. Pop accompanies her, but as soon as she begins incubating, he swims off to the southeast section of the pond. I leave and go to town to buy a kayak. I've finally got an excuse to purchase this expensive toy—I need to get close to the nest to watch and photograph upcoming events.

I have splurged on a ten-foot-long, squat, red plastic "Walden," and the moment I launch it and start to paddle out to the nest Jane lowers her neck onto the nest rim in a futile effort to hide. She waits until I come within six feet of the nest before she jumps off and honks her displeasure. Expecting Pop to heed her noisy summons and come rushing to her aid (since he has "protected" her and the nest even from a mallard drake), I anticipate having to endure a thrashing. When intruders have come in the past, he has always been quick to attack them and then she has stood up, covered her eggs by scraping debris over them from the nest edge, and joined the fray. But this time Pop is nowhere within sight or sound. As I scan the far shore looking for him, I eventually see him with his head lowered down to the water, ducking behind sedges.

Pop's was not necessarily the typical response, as I found out in subsequent tests with the same kayak at neighboring beaver bogs. A pair of nonbreeding geese at a beaver pond (along the Richmond–Hinesburg Road) was nonchalant toward me in the boat. Then I went to another pond (at Sunset Lane), where the female was incubating even as two male mallards stood asleep immediately adjacent and almost on her nest. Her mate was not in sight, until I got within ten feet of the nest. Like Jane, she rose up and jumped off the nest and raised a clamor when I paddled out to her, and in the next instant her mate came flying at me, honking, hissing, and spitting fury. I beat a hasty retreat.

May 22, 2001.

The sedges quiver and the fresh green cattail fronds wave on this dawn. A slight breeze ruffles the water surface. At odd intervals I hear sounds that resemble a plucked guitar string, the loud idiosyncratic calls of green frogs.

There is still no sign of the goslings. Jane fusses with the nest around herself, preens, and goes back to sleep. Pop is his usual self this morning. He comes to me as soon as I arrive at the pond (without kayak). I know when the last egg was laid and have calculated that the hatching should occur today, and so I wait and wait, but after five hours there are still no goslings to be seen. Anxious to examine the nest, I finally go to the house to fetch my new kayak, and as I come back and carry it down through the woods I already see Pop paddle hastily toward the opposite shore. He disappears into one of the beaver canals among sedge hummocks.

I drift out into the pond and ease the kayak up to the nest. Jane rises up and hisses as I reach over and take an egg from under her. It seems huge, and I'm almost startled by its warmth. I hold it up to my ear. Are those scratching noises? I can't be sure. But then I hear a sound—just one. A perfectly clear, loud, high-pitched *cheep.* I quickly replace the egg and she immediately sits back down on the nest. Then I paddle over to the other side of the pond to search for Pop. I cannot find him, and so I leave the pond, drag the kayak into the woods, and go into hiding behind a clump of beaked hazelnut bushes on the hillside.

After twenty minutes of watching from the woods through binoculars, I see Pop's head peeking above the sedges. Jane, too, looks for him. Her head is high and she swivels it around, then gives a single honk. He starts pacing back and forth behind the concealing sedges, his location being continually revealed by a male red-winged blackbird who keeps dive-bombing him. Until now the redwings have ignored the geese and I therefore suspect they may be alarmed by his surreptitious behavior.

Pop swims to the end of a sedge point, turns around, and swims back to keep cover between us, then stands for minutes with his head just high enough to peek over the cover of vegetation. What is he hiding from? Can it be me? It seems so, even though he allows me to touch him and even though he swims to me in the morning as soon as he sees me. His real concern is the kayak, though, and having just seen me in it, perhaps he now associates me with it. At first I think he is being not only cowardly but also stupid, because there is no big bad kayak in sight, but I reconsider when I remember that people sometimes similarly condemn by association and conclude by correlation.

After he has been in hiding for forty minutes, apparently from me standing in the woods, I try to hide myself even better. This does the trick: He slowly eases out into the open water of the pond and starts to cross toward the nest, swimming very, very slowly with his head high. Jane, on the other hand, is by now asleep. Her head is on her back and her eyes are closed (I see the white lids with binoculars).

At 11:53 A.M. a loudly honking goose approaches and flies onto the pond. Jane rises tall in her nest and calls. Pop now honks, too. She jumps off the nest and attacks. He follows her. In seconds the new goose is off the water, making three high passes over the pond before departing west from where it came. The interloper gone, Jane paddles back to the nest and he follows her. As she hops back on and begins to preen, he rises tall and does the feel-good wing-flapping ceremony that he always performs when he has accomplished something and seems to be taking credit for it.

When I tell Rachel about Pop hiding rather than coming out to defend his mate she says, "As far as I'm concerned, he already had one strike against him" (presumably for not having waited for his mate, Peep, and taking Jane instead). Now he has maybe others—cowardice and false pride—or taking credit for chasing off the stranger. Our morals are matters of interpretation that mainly concern intent, as opposed to the actual good or harm done to others. We don't know his intent. However, geese and humans have evolved similar social systems for long-term bonding to mates and young. And we may therefore have similar standards of behavior.

Specific behavioral tendencies have evolved in geese, just as they have in humans, because they benefit self or offspring. Pop recognizes me and he may be apprised that I know where the nest is, so his absence or presence near me conceals nothing. But a strange new thing like a kayak can be dangerous if it is big and could be a sign stimulus of a predator to him, as the noisy truck had been a sign stimulus of a flock of geese to Peep. His hiding may have nothing to do with cowardice; instead it may be prudent for a parent goose to hide when a predator is near because a goose on a pond can attract attention and lead to discovery of the vulnerable young. He made no attempt to hide from the coyote because that was long before the arrival of helpless young who must forage on the shore for grass. Pop would not "know" any of this consciously, but most of us also don't know why we feel compelled to do many of the things we do automatically and without thinking.

May 23, 2001.

I came to the nest four times today, expecting to see goslings at any time. Finally on the last check, at 8 P.M., I saw a dime-sized hole in an egg and through it some wet, matted down. Each time, Pop came toward me when I was without the kayak, and hid when I came with it.

May 24, 2001.

Pop is again (or still) directly by the nest when I return (without kayak) at 5:45 A.M., as he was last evening. But today he does not budge from the spot when I call him from shore to offer grain.

Even from shore, I see an empty eggshell at the side of the nest. The top of the egg looks ragged, as if it had been chewed off. She lifts her wing and I see wiggling yellow fuzz! A raven flies over—Pop moves his head to follow its flight path over the bog.

After about fifteen minutes he suddenly leaves her and the nest and hesitatingly starts to paddle over to me. I encourage him by looking at him and talking loudly so I don't seem surreptitious or interested in anything that's going on at the nest. I say, "It's okay, Pop. Come in. Come on. . . ." Of course I don't expect him to understand the literal meaning of what I say, any more than I need to attach literal meaning to understand many of his utterances. He presumably understands my intended meaning—and that I don't pose a threat. Indeed, after a few minutes parked in the water and looking at me, he suddenly advances again and nibbles on the grain at my feet, and I reach out and touch him.

I have brought the grain because I want Pop and Jane to keep coming and bring their young, so I can see them up close. That's the idea. But logic is seldom a match for reality. Logic is never enough. It's the essence or "truth" of mathematics, but in biology it's only a tool.

After taking a brief snack, Pop paddles rapidly back toward the nest and parks in the open water about twenty feet from it. He is a father at last and seems more tense and alert. He scans in all directions. In a few minutes he leaves his position in mid-pond and returns to Jane's side at the nest. In contrast, during the last few days he has stayed far away from the nest and has often not even been in sight. Will he defend the nest now that the young have hatched? I must again bring out the kayak and test him.

As I'm unloading the kayak from the truck up on the road on the other side of the woods, I can see him through the foliage screen. He sees me, too, and leaves the nest rapidly to paddle off to the opposite side of the pond. By the time I carry the kayak through the woods and reach the pond and paddle out to the nest, he has disappeared. She, on the other hand, is only momentarily annoyed by my presence as I anchor the kayak in mud and cattails. From my position next to the nest, I hear faint peeps, and soon see little yellow fuzz-heads pop up out of her feathers between the "elbow" of one wing. She purrs and makes low, barely audible grunts that are on the lower frequency threshold of my hearing. They sound like a passel of contentment. I tell her, "I'm not going to hurt your babies. I just want to see them—it's okay." She understands, because she responds to my voice by standing up and revealing her brood. Five lively olive-and-lemon-yellow goslings tumble about. These animated little fuzzballs have faces of pure bright yellow, olive green backs, round fuzzy heads, olive green bills and feet, and dark eyes. They show pink mouth linings when they yawn. One rolls off the edge of the nest but catches itself to scramble back up. As Jane sits back down, she lifts her wing to admit the wayward one; then she closes it and the youngster is tucked in. I'm amazed that she trusts me enough to so nonchalantly expose her young to me. We have by now, however, a long-standing acquaintance that spans years.

Anchoring my kayak in the muck at the edge of the nest, I sit back in bliss and four hours pass like an instant. Occasionally I maneuver to another spot, for another view. She sometimes gives a faint hiss if I inadvertently nudge the nest with the prow of the kayak; then she again relaxes. Pop stays away entirely, and I do not see or hear any sign of him.

In the three hours till 10:30 A.M., there were three episodes, each lasting two to three minutes, when she got up and the goslings played. After each playtime they crawled back under her, and all then soon stopped their peeping. One time when she arose, she reached over with her bill and rolled the empty eggshell (which looked like a whole egg except for a large hole at one end) at the edge of the nest back under her and then sat down on it. It looked as though she confused it with a whole egg that had not yet hatched.

Seeing her deliberately rolling the empty eggshell under her, I would have taken it as proof of how stupid geese are, if I had known nothing else about geese. Imagine going to the trouble to retrieve and then incubate an empty eggshell! The famous experiment by the Dutch ethologist Nikolaas Tinbergen showing gulls retrieving huge fake eggs and taking them into their nests is taken as proof of the blind, unknowing instinct designed to incubate a *misplaced egg.* All researchers, and certainly all students (if they pass a course in animal behavior) believe it. It's in all the biology textbooks. But there may be more to such behavior.

The raven who has been flying by the pond several times each day suggests a twist to the conventional interpretation of this instinct. While I was sitting next to the nest, that big eggshell looked conspicuous, and not unlike a whole egg. I put myself into the raven's place, and when I did I realized that the "egg" was like a red flag that no predatory scavenger would miss, even from a long distance. Ravens and many other scavengers are inordinately fond of eggs—and young hatchlings. If one came down for a closer look it would discover the real eggs, or the goslings. I was glad when Jane rolled the empty eggshell into the nest, covered it, and thus probably inadvertently crushed it. (She had disposed of her four other empty eggshells by crushing them also, as I determined later by examining the nest contents.)

It is not necessary to presume that she thought about disposing of the nesting evidence to avoid predators. She might just have felt uncomfortable seeing the eggshell next to the nest, and reacted. But instead of viewing her behavior as a blind instinct to retrieve misplaced eggs without distinguishing between a real or a fake egg, or an empty dry shell, I now saw it also as elegant behavior that helps to ensure the protection of her young. Protection that might not occur if it were left to the complex vagaries of experience, learning, or remembering. One or another of the former can always be inadequate, making the rationality slanted, limited, or dangerous. Rationality might have informed her that the round, empty object was an empty shell, and therefore of no use and thus not worth retrieving.

Jane is tranquil and unperturbed by me. She starts to nap, closing one eye, or both at the same time (as I can see on rare occasions when her bill is toward me). Her neck slowly arches back every time her eyes close, in a gesture similar to when our head slumps forward and our eyes close. Eventually her nodding head goes all the way onto her back and it rests there and her eyes stay closed. I'm only five feet from her. I feel bonded to her by her trust, and I am thankful for Pop's reticence to attack.

Her catnaps are short, and I doubt that she sleeps deeply. She may look like she's sleeping when she has her eyes closed, but occasionally I still hear soft, low purring grunts of which she gives several in quick succession, with bouts separated by about one-second intervals. I mimic her soft, throaty, repetitive grunts, which the goslings always respond to with peeps. She relaxes, pulling her neck down deeper into her back feathers.

By late morning the first bullfrog and green frog chorus erupts. With it, I usually hear splashing as well, and my eye catches a nearby bullfrog. I see his lemon-yellow throat pulse with the *garumph* sound he makes, and I next see him make successive leaps toward another frog. Grackles fly to their nests with their bills jammed full of the soft, just-emerged dragonflies, then leave carrying instead white fecal sacs freshly delivered to them by their just-fed young. They drop these membrane-wrapped packets with a splash into the water, with its mirrorlike surface. There is not a drop of fecal matter in or near their nest. Like the geese, they minimize nest-indicating evidence.

Three goslings are lying down on the nest in front of Jane. One has its eyes closed and a leg stretched way back. The sleeping one starts to preen, another picks at loose down feathers, and then both go back to sleep. The third pokes Jane with its bill, trying to lift her wing to get under it. She gets up, turns her back to me, and the crowd of five goslings churns under her. She settles back down, and they poke their heads out along her side to then again come out for another tumbling frolic.

This year's crop of tiger swallowtail butterflies have just emerged from their overwintered pupae to search for mates. One, presumably a male, flutters over the goslings. It is attracted to their yellow color (which is the same hue as the butterfly's) because butterflies search for mates of their appropriate species using color as a long-distance cue. The just-hatched goslings snap at the butterfly hovering about them. The chicks miss. Jane doesn't and she immediately swallows her prey.

At 11:29 A.M., Jane rises up for the fourth time, stretches first her right leg and then her right wing, then stands on one leg. The fuzzballs open their eyes and yawn, peep, and tumble about. One balances on one foot and scratches under its chin with the other. Another rises tall on its olive green feet and legs, and rapidly flaps its wing stubs in what looks like a perfect rendition of the adult's feel-good ceremony. Others shake either their head or their tail end or both in succession, stretch out a leg or a wing, or preen their back, sides, belly, and undertail with their tiny bills. They waddle about on or just over the nest edge. One starts to pick at down, another at a piece of dry cattail. They pick at almost everything in front of them—down, nest debris, and their mother's feathers. They must soon begin to feed, and now, in play, they poke at all kinds of objects. They are learning, and their experience will soon allow them to sort out the edible from the inedible. Intelligence about the good things in life is less a matter of rational analysis than of having made many "mistakes" and of learning, experiencing. And here I speak from the authority of personal experience as well as from detailed studies with juvenile ravens, who, when leaving the nest, behave a lot like the goslings.

Slowly, deliberately, Jane settles back down onto her nest, and most of her babies again crowd under her. Three are off to the side, and she lifts her wing toward them as though beckoning them to come; they do come and rush under it. I'm starting to wonder what will induce them to jump into the water and how they will react to it the first time.

By noon Jane has become visibly more nervous. Her grunts are of longer duration, rather than being quickly repetitive. She keeps raising her head high and looking all around. I suspect she is looking for Pop, and in her apparent concern about his absence she even becomes inattentive to her goslings. One tries to get under her wing and she keeps looking away, not responding to let it in. Meanwhile the little ones are putting everything in sight into their bills and "mouthing" it.

She is staring at something on the far side of the pond. I follow her gaze. And as soon as my attention is directed by hers, I see him! Pop is finally visible at the edge of the sedges. He is alert, with his head held high. So I start looking at him, too, and he immediately responds. He puts his head back down and lays low behind the sedges. It's as though he can tell from my behavior when I can see him and when not.

As she stares at him, she makes growling grunts. Why doesn't she *call* him if she wants him? Why doesn't he call to her? Their overall behavior fits with my idea that they are acting to avoid drawing attention to themselves, because at this time of year, after the nonbreeders fly north to molt, the presence of an adult goose usually implies the presence of goslings. The two are inseparable, and predators would learn of the association. But I'm already at the nest, and the rigidity of his simplistic response is as puzzling as it is inappropriate in the present context.

With Pop out of sight again, she sits back down and the goslings again rest. There is not a peep out of any of them for fifteen minutes as she catnaps with her head laid onto her back. Gradually Pop again emerges from hiding. She is in plain view and he must know that she is totally at ease near me. Yet, he remains on high alert for predators and still tries to conceal himself from me, from whom he has nothing to fear.

At 12:16 P.M. there is again a stirring by the little ones. Two pop their heads out onto her back from under one wing. Later they come out and waddle next to her, picking at her feathers, her tail, wings, and cattails indiscriminately. One of them defecates yellowish liquid feces even though I have not yet seen them eat or drink. She yawns, and at 12:25 she finally rises and stretches both wings up at the same time. She sits back down but keeps her head up, to keep looking all around. Soon she rises again and the goslings tumble out from under her, yawn, stretch, and again pick at everything in sight. Another voids a stream of feces. Jane has become restless and impatient. She keeps looking all around, paying scant attention to them. After standing for 24 minutes she sits back down, quickly rises again, and stretches, preens, then sleeps. But in only 4 minutes she is again up and on high alert, then sits down to sleep again, and at 1:01 stands up once more. As before her attention is directed toward where Pop is hiding.

Five minutes later, after briefly sitting down, she rises again, stretches her right wing and leg, stands on her left leg, preens, looks around, and preens her belly feathers. Now, *finally*, Pop slowly and cautiously starts coming across the pond toward us. Seeing him come, she immediately stands even taller, following him with her gaze. I look, too. As before, the second or two after I catch Pop's eye he swims back to conceal himself behind the sedge peninsula he has just left. So it's not just my presence that's critical to him—what matters most is that *I* notice him (and that I'm with the kayak). Apparently, he knows the significance of eyes.

Jane and the goslings have probably long since been ready to leave the nest, but they haven't because Pop wasn't with them to accompany and protect them. They need his protection. I now confidently predict that they will leave as soon as I go, because Pop will take the first opportunity to come take them away.

I test my hunch by paddling to shore, then running up to the house (to grab a piece of bread to eat). In two minutes I'm back down by the pond. As I had expected (but not so quickly), he is *already* directly beside the nest, and she and all the babies are in the water next to him. Entering my kayak, I paddle back out to them. Will he now defend the young when I try to photograph the mother with the goslings in the water?

As I launch the kayak onto the pond, she makes haste to swim off along the edge of the cattails. The young at first look as though they are attached to each other in one thick knot. But they string out behind her when she swims fast, and they all instantly reaggregate into a tight bunch by her head when she stops or slows.

They are fast and "expert" swimmers who bob like corks on the water. Pop hangs back although he now neither hides nor attacks.

I expected Jane and the young to follow Pop. Instead, as I approach she returns toward the nest: Pop remains conspicuously parked out in the pond and watches from a distance. But the instant Jane gets near the *nest*, he again vanishes into the sedges. Apparently, the evolutionary logic of his behavior is to keep the location of that *nest* a secret. Unfortunately, he does not know the objective behind the behavior that drives him. He is confined to operating according to simple rules laid out by his evolutionary script and is missing the instructions applicable to this (new) situation.

And as I anchor my kayak next to the nest, instead of retreating with her goslings as she has done only seconds before, Jane now leaves her young and comes toward me and hisses. She appears to be *protecting* the nest. She, too, has not yet recognized the changed reality. Heeding her threats, I draw back slightly and she instantly hops onto the nest. One after another of the young then also clamber up, and she settles down to brood them, presumably for protective cover rather than heat.

Again I'm inclined to think, "What a stupid goose—she deliberately comes back to defend the *nest*, rather than defending the young when I chase them." How illogical! She has protected the nest before, and still does, as though not realizing the nest as such is no longer of value. However, maybe I am the one who has it all wrong. There is another perspective: She learned from long experience with me that she and her young are safe at the nest. Nobody hurt her *there* even when potential dangers were near. Her babies would again be safe there. It's a logical conclusion, but not an intelligent one.

Jane and the family waited for Pop before leaving the nest. They should do so again. To find out, I leave the nest and again paddle to shore to see if he will return. I'm barely ashore when I already see him hurry over. So I stop and hide to watch developments (after dragging the kayak out of the woods and back up to the road), and I experiment: Three times in a row, the second I stop to *look* (even from afar through a thick screen of foliage and at sufficient range to require the use of my binoculars), he stops, turns tail, and goes back into hiding. He, who previously fed from my hand, who let me touch him, and who usually came to my call—now retreats when he sees me *look*, from fifty yards away.

I then leave for an hour. On my return I find him tall and erect, standing on the nest. He looks jerkily all around, and when he sees me, he instantly jumps off, puts his head down low, and paddles off putting a patch of cattails between us. Soon I see two heads; Jane and the goslings have not gone far, and he has joined them.

As the family scoots behind an island of sedge, I train my binoculars on the other side of the island hoping to see them when they emerge into view there. She and the goslings do appear. But he stops short of the end, lets them go, and turns around to again keep cover between us. Jane and the goslings eventually return to him when he doesn't follow them and then mill around him. Perhaps Jane realizes something is "wrong" with that area, maybe not knowing that Pop's unease comes from my eyes, far away, not from a danger there. Finally, she and the goslings venture out and leave him, to hurriedly paddle across the whole open expanse of pond all the way back to the nest, her safe haven. He follows. She again hops onto the nest, and as before the little ones waddle up and join her.

He now stays about five feet from her and the nest. And then he does another extraordinary thing. He repeatedly makes the "Let's go" signal—the rapid head-jerking motions. This time the gesture, one I have seen hundreds of times, is not accompanied by the loud vocal display that previously has always accompanied it; he remains totally silent. He obviously wants to take the family someplace, I know not where, and he is not willing to advertise the journey.

9. TO GREENER PASTURES

May 25, 2001.

No geese are in sight at dawn. They have left the nest, and only a rotten egg remains. I call for Pop and toss grain in case he may be watching from hiding. The redwings and grackles continue their errands and an oriole and catbirds call, but I scarcely notice as I strain my senses and still do not find the geese. Could they be at the upper pond? The beavers have left and the water is drained due to the slow leak in the dam, but fresh grass is starting to grow on the dam and on the increasingly exposed pond bottom.

I cross through the woods and when I come out at the edge of the upper pond, I see Pop standing next to the beaver lodge with Jane squatting down nearby. Her wings are partially spread out to the sides. The goslings, who are presumably being warmed by her body and shielded from sun and rain by her wings, are not visible. I approach the family across the old leaky beaver dam that leads directly to them: "Hey Pop, Jane, I'm coming over to see you and the kids. It's okay. Just relax." I could of course recite random numbers, but then I would not be conveying my emotions and would also be acting oddly.

Today he doesn't try to hide or run.

Even my gaze doesn't seem to faze him. His head remains high, and as I draw closer I continue talking and acting uninterested in him by looking down or out over the bog at the blackbirds. Neither Pop nor Jane budges as I scatter corn on the bare banks of the dam, and I walk back into the woods to watch them.

Jane gets up and five lively goslings scoot out from under her and follow her as she and Pop wade into the one remaining shallow pool of this beaver pond. Corn is no longer on their menu—the family paddles across the inlet channel to reach a mudflat on the other side where grass, suitable for baby food, is sprouting. Jane leads and Pop brings up the rear. The goslings scramble onto ground and some immediately start grazing; others rise tall on their tiny feet, beating their stubby down-covered wings just like the adults do. I'm elated to see them at ease and I hope to see them now every day as they grow up to adulthood.

May 28, 2001.

I found no sign of my geese yesterday, and checking four times, I still found no trace of them today. I searched the whole periphery of both ponds by walking back and forth through the sedges. On reflection, the disappearance was no surprise. The mystery was their whereabouts. Geese have nested here previously and I never saw the families afterward for more than a day or two. I just didn't give the disappearances much thought because I presumed I had not checked carefully enough. But now their absence was real, and since the families had vanished already several times in the past, the disappearance was worth investigating. As a first possibility of what might have happened, I tried to find out if they ended up at one of the several neighborhood ponds that are found within a radius of about five miles of our pond.

Surprise is hardly an adequate word to describe my feelings about what I found: Eventually all seven of "my" geese were located two and a half miles (four km) away at a small farm pond at Sunset Lane (see the map across the page). The woods in between our beaver pond and the Sunset Lane pond presumably contain the usual fauna of coyotes, foxes, raccoons, cats, dogs, skunks, weasels.... Could they have run that gauntlet? If so, why

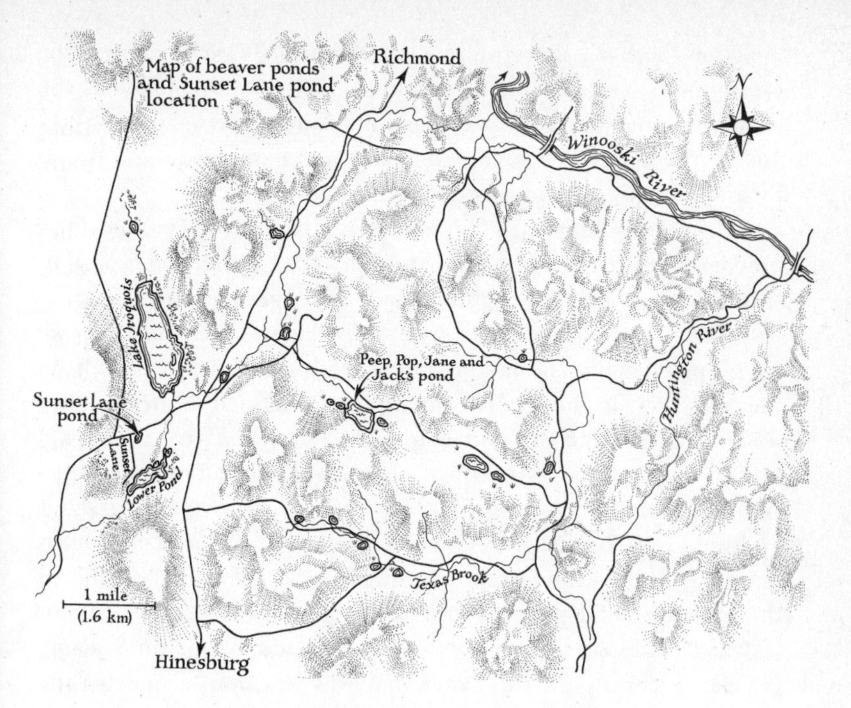

A map of the beaver ponds and their locations.

would they have risked it? I had once surprised a woodcock with her week-old chicks, and the bird, to my great surprise and wonder, flew off carrying one, or maybe two, wedged between her thighs and/or in her feet. Waterfowl are not capable of such feats.

Unlike the beavers' ponds, Sunset Lane pond is surrounded by hayfields where cattle graze. To reach it the goslings had to walk not only through a half mile of forest, over a hill through hayfields, but also across a road. Rachel had told me of seeing a family of geese crossing the road into the hayfields on the top of this hill the previous year. They were heading in the direction of Sunset Lane pond. But that pond seemed to be an impossibly distant target, and the improbable wasn't at the level of the potentially possible in my mind at that time. Rachel's observation was puzzling and interesting, but I had no context for it and so did not know what to make of it. Now the observation assumed significance. Last year (2000) a pair of geese had a nest on a small island near the southern shore of this (tiny) Sunset Lane pond, and the family of two adults and five young foraged in the fields at the edge of the pond all summer long. I was later surprised to see a second family with three young there also, and I assumed the simplest hypothesis—that I had missed their nest.

This year I was certain that only one pair (probably of first-time breeders) had made a breeding attempt at that pond, and typical of young geese, their nest contained only four eggs and the pair incubated them only sporadically. After they had not incubated for two days, I paddled out and found piles of fresh goose guano directly on the cold, now dead eggs. The young pair remained at the pond for a while, even after Pop and Jane came with their five goslings. I saw no conflict. The two pairs of geese were tolerant of each other. Indeed, two days later, on May 31, a *third* pair arrived with four goslings. This family grazed at the pond edge within a few yards of Pop and Jane and their family. Finally, a lone goose joined the growing pond population and this goose was also left unmolested. I could see no interactions. Seven adults and nine young grazed peacefully on this pond that was less than a quarter the size of my beaver bog complex where the geese had a month previously, and every year, battled fiercely for exclusive occupancy. These bizarre and complex goings-on are unexpected in an animal not popularly credited with a great deal of sophistication.

By June 6 there were still seven adults—the two pairs with young, the pair without young (who had deserted their four eggs), and the single adult. All stayed close together, sharing the nine young among them to produce one big family.

I left for four days, and when I came back, on June 10, I thought I was hallucinating. I checked and rechecked. No doubt about it. There were now only four adults at the Sunset Lane pond. The three nonbreeders had left. Pop and Jane and the newly arrived pair with four young were still there. So far so good. The inexplicable part that made me think I was "seeing things" was that Jane and Pop's family had increased from five to ten young! I don't believe in spontaneous generation, and the alternative hypothesis I was left with, namely, that a fourth family had come and left their five young off with Pop and Jane and then left, was at face value also not credible. Any logic I might apply undoubtedly would be premature. I needed more observations for a possible explanation.

No new geese arrived over the next two months. The three non-breeders that had left stayed away, but the two pairs with goslings remained. The ten young stayed with Jane and Pop the whole summer, as did the four with the other pair. The two pairs sometimes mingled as they grazed, even as the young of each pair stayed closest with their own parents, adopted or otherwise. In early August when the fourteen young had feathered out, a casual observer would have seen merely one flock of eighteen geese. He would probably not have suspected that *none* of this flock of fourteen young had hatched at this pond where they spent all summer, that the original pair who had built a nest at the pond and laid a clutch of eggs in it later aborted their breeding attempt and then left, and that the total of eighteen geese consisted of two families, one of which had five adopted young in addition to their own five that had come from a nest on a beaver lodge far distant from this place.

Pop and Jane had been model parents. They eventually raised their five young to healthy adults who left the pond after they had grown up in late August. Their adoption of five additional young had cost them little, if anything, since geese don't gather food for their young and since the food supply in this and surrounding ponds is virtually endless. Jane and Pop's main role as parents had been to lead their young to this pond and to serve as protectors from predators such as the locally common coyotes and foxes. Maybe by adopting others into their brood, Pop and Jane had in effect acquired a predator shield; the larger the flock, the more eyes to see danger and the smaller the possibility that if one of the young should get caught it would not be one of their own.

Two Shelburne pond pairs
Pair A
Pair B
♂
♂
♀
♀

Heinsburg-Richmond pond pair
(right and left profiles shown)
♂
♀

I learned to recognize individual geese by their unique facial markings. These sketches of several pairs, all drawn in 2002, show examples of the identifying variations I found.

10

A SUMMER NIGHT CONCERT: SOUNDS AND SIGHTS

In early June the beaver pond that the geese have just left is surrounded by a lush green wall of foliage. Cattails, blue-flowering pickerel weed, and sedges encroach in the shallow water from the sides, and the pond surface becomes layered with brown-green pondweed and sprinkled with both white- and yellow-blossomed pond lilies. Showy blue flag irises flower in clumps among the sedges, and nannyberry bushes along the shore and on the old beaver dam erupt with panicles of tiny white florets. Swarms of male blackflies no longer dance up and down in unison over the water surface in their communal mating rituals. But their now-mated females, who earlier emerged from streams, are in hell-bent hordes hungry for blood meals to mature their ovaries and produce eggs. The geese have by then led their young away from this pond, but numerous other birds remain and they are busily nesting. The dawn bird chorus has become muted, and the raucous day and night chorales of thousands of wood frogs have long been silent, while those of the spring peepers are subsiding. A green pall has set in. The animal exuberance of early spring seems squelched as the mating season is submerged by the demands of rearing young. This is not my favorite time of year. I like animal noise, and one night I unexpectedly heard it.

I awoke on the night of June 10, 2001, hearing deep resonant tones like those from the plucking of giant bass fiddle strings attached to an electronic sound synthesizer with an amplifier. These were bullfrogs. I often hear bullfrogs, but only as individuals. Tonight, when one frog bellows, several join in, and that sets hundreds of them off in a continuous bedlam that lasts 8 to 10 seconds and then stops abruptly. For 10 to 12 seconds the silence is total. Then the fiendish cacophony resumes. The pattern is endlessly repeated. A Silurian swamp could have sounded like this, I think, as I lean out the bedroom window and listen to the beat. This is a concert I cannot miss; I have to get up and go down to the bog to experience it.

There is no wind and sounds carry far. Stars are out but no moon is yet visible at 11:25 P.M. Occasionally lampyrid beetles (fireflies) flash in the dense maple foliage above my head. I walk in the darkness through the woods and down to the bog and think of the tropical lampyrids who gather in huge crowds and signal (flash) in synchrony to produce a much greater signal than one can alone.

Bullfrogs commonly perch among the pondweed in the daytime at the upper end of the pond where I sit near the goose nest. However, tonight all their commotion comes from the opposite end, near the dam. Not one bullfrog is calling from where I have seen them during the day. Apparently they have left their usual haunts to gather up into a crowd by the beaver dam where there is less pondweed and the water is deeper.

During the bullfrogs' regularly repeated 10 to 12 seconds of silence, I hear a lone tree frog trill a mile away. One lone spring peeper still calls although occasionally a second one chimes in. At random times, and seemingly random places, I hear a quick energetic *gunk*, the plucked guitar-string plunk of a green frog, or the snore-call of the pickerel frog, a sound easy to miss. Intermittent splashes fill the silent spaces of the unearthly sounding bullfrog chorus, and the combination of sounds coming from the fetid darkness on this sweltering night beckons evermore to thoughts of the Silurian.

I asked two well-known frog researchers, who have spent

decades trying to decode frog vocalizations, what was going on. Rick Howard from Purdue University said he had similarly heard bullfrogs pulsing, and he wondered if possibly the males were "listening for something between chorus bursts." He had determined that the females seemed to know when all the gravid females had mated, after which the males stopped calling. Kent Wells at the University of Connecticut speculated that when males are close to other males they may match their own calls to their neighbors' for the attention of females, but maybe they can't keep it up for long and so fall silent to rest. These are potent considerations. However, woodfrogs also join in when they hear others (I tested it with playbacks of taped choruses), and they produce no *pulse*-chorusing like that of the bullfrogs I was observing. Instead, woodfrogs sustain a long, continuous bedlam, although there are occasional breaks in it. In some amphibians (spring peepers and some toads) there are "satellite" males who don't call at all, but instead parasitize the calling efforts of others to attract females, thereby saving their energy for mating. The denser the aggregation of calling males, the greater the tendency for males to remain silent. Indeed, in some anurans (frogs and toads) that breed explosively in dense aggregations (such as the European common toad, *Bufo bufo)* the males have lost vocal signaling. Clearly, the bullfrogs' behavior contradicted that model. They joined in when others called, rather than being silent. Therefore, it seemed likely to me that for them and woodfrogs the *louder* overall chorusing produced by joining in with other males has some specific function. Does it serve to draw in more females from a wider area?

These bullfrogs were presumably males calling to attract females, and I'm puzzled why they aggregated and organized their sexual signal into a chorus rather than "going it alone." Is theirs yet another example of animals using cooperation to compete against one another? Bullfrogs wander—I've found their carcasses as common roadkill. It is safe to assume that a louder or bigger signal can attract a female from a farther distance. If a frog can attract a female from half a mile away by using his loudest bellow, then by joining with another to call in unison he should reach a female that is a mile away. Alternatively, a female should be attracted to the pond with the loudest signal because that pond will seem to be the closest. It is advantageous for her to try to reach the closest and/or loudest caller because to do so will cost her less in terms of time, energy, and risk in predation. If signal amplification via joining another caller is good, then joining fifty is even better. Of course, the downside to a frog "howling with the wolves" is that any female he helps attract to the pond may mate with any other male in the crowd and not with him. Still, if as a group the males are able to attract a larger number of females, then the chances of successful mating are greater than that of any one individual. One hypothesis does not exclude the other. Maybe joining a neighbor's calling is driven by inter-individual competition, while at the same time functioning in interpond, or group, selection. Undoubtedly it will require a few Ph.D. studies to sort it all out, and the answers will vary depending strictly on which species, populations, and habitats are investigated.

The bullfrogs' durable aliveness contrasts with the ephemeral, yellow-green flashes of the lampyrids. As the males of these predatory beetles fly silently through the trees and over the sedges, they blink on-off in flash patterns that induce females of each species to give their appropriate species-specific "answers" so that the males are lured down to mate (although some females have evolved to lure in the males of other species, to capture and eat them). One of the lampyrid beetles on the ground near my feet at the edge of the water is giving off a fluorescent glow that waxes and wanes. Another who is stationary emits an eerie light from its perch in a small clump of cattail. This beetle's light is the brightest, weirdest firefly light I've ever seen. It starts off faint, slowly gets brighter until it is startlingly luminescent like a blue-white headlight, and then just as slowly recedes. These prolonged "flashes" last about two seconds each, and there are always three in succession. I'm clueless about the specifics of most of the animal communication around me, and that extends even to the insects that seem to act like windup toys. Can I ever understand a goose?

11. FALL VISITORS

As sure as the arrival of fall colors, every September-October cold front brings flock after flock of hundreds of Canada geese flying south. Day and night they come over, and the ceaseless, passionate cries of the honkers high in the skies make me look up and feel restless. Occasionally a noisy group of fall-flying geese comes down to land in the beaver pond in the evening. The birds remain quiet in the night but near dawn they clamor to work up enthusiasm for early departure. One bird's excitement kindles the other's until through mutual stimulation a threshold is reached and wings start to slash through the shrouds of morning fog lying on the pond, and the crowd lifts off.

On this beautiful day (October 8, 2001) at the peak of the fall foliage, the day after the first snow, I'm happy to again hear geese on our beaver pond at daybreak. The crowd has probably rested there in the night and will now be departing. The calls are a powerful summons, and I get up hastily to brew a cup of coffee and watch their takeoff. With the coffee still warming my hand, I walk down to the pond. Unfortunately I'm late; the crowd has left just before I get there. Instead, I see several mallards dip for food near shore, showing their light bellies in the gray dawn as they tip over with their tails up. An otter with its head out of the water looks around, stays stationary for minutes, submerges in a smooth swirl, and comes up again. With only its nose showing, the otter then etches a V on the water surface while swimming steadily across the pond.

The now orange, crimson red, and lemon yellow of the red, or "swamp," maples on the other side of the pond are luminescent in the early morning light. Recently fallen leaves at my feet are dimpled with beads of water from the melted snow. The ducks make a quick succession of quacks, then fall silent. But at this season I'm not long without geese. In the distance I already hear more faint cackling, and soon a flock of about 100 flies over the pond. By 11 A.M., six more flocks of 100 to 200 birds each will have passed, pushed by the cold front that has brought the snow from the north. The birds fly not only when they make the most distance with the least cost but also by drafting on each other to save energy. Each individual in the flock is slightly to the side and behind the other to form a V that cleaves the air and aerodynamically reduces drag and saves energy to all but the leading bird. The two long arms each trail back from the sharp apex, like two thick ropes flying through the sky. I watch a bend in the beginning of one arm of the V ripple slowly all the way to its end as each goose, one after another, realigns in sequence. Geese at the end of one line may detach, then reform into smaller groups, one behind the other, and throughout, they never observe a moment of silence as they smoothly wing along on their way south.

On October 12, I again hear goose clamor in our bog. But this time I hear only one goose calling, and it is making a long, anxious summons. I listen more closely than normally, and more interestedly, but as I'm about to jog down the driveway the goose has already taken the initiative. It begins to call more loudly to prepare for takeoff, and I expect it to be gone in seconds. What a surprise, then, to hear it come closer and to then see it circle the house. The goose is flying all alone. "Peep!—Hello Peep," I holler on impulse.

The bird swivels her head and comes in for another flyby. I call again. She makes one pass after another over me—then, finally, she starts a glide as if to land, only to resume wing-beats and regain altitude at the moment when she has almost threaded down into the clearing by the house. I crouch down onto the lawn and call again, and she then comes again, sets her wings, glides down, extends her feet, and flutters to a landing near me. It *is* Peep! Meanwhile, Eliot has come out of the house, and Peep slowly but unhesitatingly walks to us. We pull up grass, and she takes a few blades from our hands, even as the whole lawn is available and she has walked over that lawn to come to us. I get out the corn for a special treat, but she accepts only a few kernels as she continues to make her low grunts and also her little purr. She lets us stroke her on the back.

I then hear another goose—just one call—down by the pond. Who was *that?* She must have left other(s) behind in order to come to us. I'm curious to find out who was with her and want to go down to check, but every time I start to walk down the driveway with Eliot, she waddles at our heels. I do not yet want her to leave us, as I know she will if I go all the way to the pond. So we return to bring her back. But finally it is time to go, and Eliot, Peep, and I walk down the whole length of the long driveway through the woods together. Her waddle is slower than our walk, and when we get about twenty feet ahead of her, she flies to catch up. After passing through the woods, the three of us follow our dirt road a short way, then again walk through the next patch of woods down to the pond. I immediately see a pied-billed grebe, a great blue heron, two mallard drakes, and a hen. And then on the far side of the pond, we behold a tightly bunched group of four geese. Peep slides into the water, calls once, and then lifts off and flies over to the companions she left less than a half hour earlier in order to come to the house.

For the next half hour Peep no longer responded to our calls as she stayed with her fellows, and an hour later the five lifted off and left the pond together. I did not know where this group had come from or where they went, but they wouldn't be back. They were not tenants of this bog.

It was a brief but, for me, significant encounter. After months of absence, Peep and I exchanged important information: We were okay, and we were glad to see each other.

As Eliot and I walked back up to the house, he said, "I miss Peep already." I did, too. I doubted that I would ever see her again. But in my mind I would long fly with her.

October 19, 2001.

I hear geese off and on all night at the bog. It is cool (40°F), windy, and still "totally" dark at 6:30 A.M. when I go down, and at first I don't see them. Robins call, and several mourning doves fly by at the first faint glimmer of dawn. I hear the *tseep* of a few sparrows. But soon the geese start to call back and forth. Are they those dark spots out where the wind ruffles the water on the other side by the beaver lodge? Yes—they move, and there are many.

By 6:50 A.M. it is beginning to get light. A muskrat returns to its lodge, which has just been rebuilt (after being flattened by the goose nest last spring). Three mallards quack, rise, flap their wings, and settle back into the water. Then they take off and fly several high circles over the pond, as if trying to see the sun rise. Apparently satisfied, they spiral back and splash down.

At 7:00 A.M. the geese—I can count twenty-three—are milling about and getting noisier. They don't feed and seem impatient yet reluctant to lift off, as if waiting for a signal. Then, at 7:22 A.M.—it's light now—one goose in the lead of a long line of them that is facing the wind starts vigorous headshakes: they're ready to go. A leader says so. A vocal assent accompanied by numerous headshakes ensues down the line, the leader responds to them by calling even louder, and finally those in the rear are the first to take wing. Without even circling, the throng wings straight down the valley.

Another big flock landed in the pond at dusk. I watched them until the sun went down, and as it got dark I noticed a group of four snipe rise out of the bog, circle, and presumably continue on their migration. A fifth snipe then came up also, circled, and flew off alone.

October 23, 2001.

I am back this morning at sunrise. The leaves are almost all down and there is dew on the browned sedges. The mood-enhancing smell of decaying leaves is in the air. Droplets of water hang from the *Viburnum* twigs. The geese, silent and motionless, are scattered into groups. I hear the excited quacking of mallards and a beaver crunching on the bark of a branch it has dragged into the cattails. At 6:50 A.M. the first goose begins to call. A second, with a slightly higher pitch, chimes in. And then their two voices reverberate over the pond and echo from the hills all around. It gets light quickly. More geese chime in. They mill around and raise their bodies out of the water to wing-flap. Then they duck their heads down to throw water onto their backs, occasionally play-chase, and call more and more. Three individuals finally rise up and fly down the valley. Others then make higher and faster calls, and in seconds the rest (all twenty-nine) rise almost at once and stream down the same valley after the first three.

Silence resumes. But it is a perceived silence that stems from the absence of the noisy geese. After a while I sense what I had not heard before. I notice—but only after conscious effort—probably the last of this year's crop of crickets and katydids. One cricket species buzzes steadily, another chirps three times in quick succession, and still a third generates a steady pulsing sound. Spring peepers, who left the pond in the spring, now occasionally sound off from the surrounding woods. They make a different, almost unrecognizable, sound compared to their spring chorusing. Suddenly, as the dawn approaches, I hear a whistling-swishing sound overhead—four mallards are zooming in. Then three more come from another direction to all splash down on the pond. None give their long quacks this morning but instead opt for rapid staccato cackles.

November 9, 2001.

Going down to the pond at dawn has become a habit for me. It feels right. This morning I listen to the wind blowing through the bare trees as dark clouds drift over and I see the pond surface alive with white reflections of light that twinkle off the wavelets whipped up by the wind. I can't see the wind any more than I can see genius without works, but I see its course as it charges and makes swirls on the water. The animals and the bog, too, are like random ripples that conform to or combine into a larger picture, if one can only detect it.

A great blue heron rises from the near shore, flies ponderously to the other side, stretches its wings for a short glide, and comes to rest at the shoreline. The leaves are lightly dusted with tiny granular snowballs, and the snow, driven by a cold, brisk wind, makes a tinkling sound on the leaves at my feet. A lone blue jay flies by silently, showing brilliant blue against the early winter pastels of tan, brown, and reddish twigs. The jay lands on the ground by the browned sedges at the edge of the pond. Then a raven wings near, *quorks* while traversing the pond, then continues as silently as it approached while buffeted by the wind and dipping its wing in a playful twirl.

It is too cold for the frogs to be out, so a little *plop* in the water near my feet draws my eye. I see tiny bubbles rising to the surface of the dark water as a small, black, mouselike shape scrambles under the water along the edge of a sedge tussock and briefly flashes white before vanishing. I think I've just seen the back, and then the belly, of a water shrew! If so, it is the first time I have ever encountered one alive.

November 12, 2001.

There have been ever-stronger hints of winter. This morning it has arrived. The leaf-strewn ground and naked tree branches are plastered with fresh, white cushions and more snow is falling in the still-dark and quiet dawn. I love hearing the dry rustling sound of the winter's first snowflakes pummeling my jacket and the *tseeps* of chickadees, who are early risers like me. The muskrat and beaver lodges already stand outlined in white in the semidarkness, and the beaver dam shows up as a white line against the pines in dark silhouette beyond. Pine branches, twigs, and the edges of the cattail leaves all show up as sharply etched dark lines against their topping of snow.

Yesterday the clouds were coming from the north. That's fall migration weather. But the geese that had recently been flying high and making their haunting cries have long passed. Now, in the dark dawn with its gently falling snow, I hear instead the voice of a migrant who comes here from the Arctic tundra: A snow bunting, normally a flock bird, repeatedly makes a *trrrt-treet* sound as it circles the pond. Perhaps landmarks are blurred, and the finch is lost in the storm or is temporarily arrested by the gray patch of pond it sees below. It passes on and silence resumes.

I become distracted by a blue jay who flits by me with an apple in its bill. An empty-billed companion follows. At 6:50 A.M., as it is getting light, the by-now-familiar raven again comes from the east, flies over the pond, calls a few times, and continues on in silence. The chickadees can still be heard faintly above the rustle of the snow pounding my jacket.

By the next day the pond was sealed with ice and covered with snow. Peep had not been back since her visit on October 12, a month ago, and I no longer entertained prospects of seeing her anytime soon.

12 SPRING SORTING-OUT

Peep's short visit on October 12 had opened new possibilities. I now knew she was still alive, she was still tame, she knew her way back, and she might return again if she survived the winter. She had been with four others, one of whom had mildly rebuked her when she came back to them after her short visit with me. Was it a jealous mate? Was she ostracized? With these thoughts in mind, I eagerly awaited the spring of 2002, wondering if I might see her and if and how she might adjust.

Spring started early. The first flock of male redwings returned on February 26. I expected that there would still be about a month of snowstorms and heavy frosts before I would see any geese. However, two weeks later, on March 12, I saw the first honkers fly over the still solidly frozen pond. Two days later a pair landed. These two were new and unfamiliar to me. The female had a distinctively notched and narrow facial mask while her gander's white face patch was rounded and regular, unlike Pop's uniquely hatchet-shaped one. Because they took an early interest in the eastern area of the bog where there are extensive patches of sedge, I named these newcomers "the Sedge pair."

As it was getting dark (March 14), I saw the Sedge pair walking on the ice on the pond. I also heard a coyote chorus, and another answered from a nearby hill. But coyotes are common here. I expected to see the geese the next morning and came at dawn.

The Sedge pair was gone. A mourning dove was cooing owllike in the semidarkness—*who . . . whooo . . . who . . . who . . . who.* Male redwings soon sang *oog-la-eee,* and in the distance I heard the melodious liquid song of a robin.

It soon became obvious why the geese were gone. When it got lighter, I saw, far out on the ice, the freshly mangled body of a deer who had probably been deliberately chased there by the coyotes in the night. The commotion of coyotes killing and dismembering a deer on the ice (a well-known tactic of local coyotes) is not something the geese would have stayed around to witness. I felt saddened by the suffering of the deer. But I did not dwell on it. After all, except for those who are shot, that's pretty much how all deer die. Normally they all get ripped down and devoured after they weaken or make a mistake. The robin still sings. The redwings yodel. Young ravens grow.

I hid in the woods and leaned against my favorite large sugar maple tree to see what would happen when the local raven pair that nest on a small cliff near my house flew by. The female would soon be incubating. The deer was presumably already being eaten by coyote pups, and hopefully some of it would soon reincarnate into ravens as well.

I watched for an hour and a half. This time no raven came by. Blue jays strafed through the *Viburnum* thickets near the edge of the pond, but they seemed oblivious to the deer, although its meat would have been nourishing to them. However, a crow did see it, and for minutes it let loose with successions of loud staccato *ca-ca-ca-ca* calls. Its mate joined in. One of the pair flew over the deer and back to the pines, and then both left together.

In three days all that remained of the deer on the ice was a mere blotch of blood and fur as the coyotes returned nightly to feed. In those three days the Sedge pair stayed away from the bog. However, a lone goose came, landed on the beaver lodge (too far away for me to identify her), called loudly, and looked all around. After minutes of anxious calling, the lone bird made the "Let's go" head-bobbing signals, honked some more, and not needing an answer left. Like the redwing males, who at first only return to the bog for an hour or two on sunny days and then gradually come more often and stay longer each day, the geese at first commonly stay only for part of the day, until they remain full-time after they finally nest.

The Sedge pair returned the day after the deer was eaten, and over the next two weeks as the bog was coming to life, my diary entries indicate, they were preparing to nest.

> It is still almost dark at 6:30 A.M. (on March 30, 2002) with a thick overcast of swirling, restless, dark clouds that are sweeping up from the south. Powerfully and inexorably like the tide, they are bringing in the spring. Near shore the ice is already melted all the way to the muskrat lodge and the dark water is agitated by silver riffles from the wind. Patches of riffles and small wavelets advance and retreat with each gust, but here and there V-like wave patterns erupt on the water, as fish near the surface zag erratically.
>
> The Sedge pair loiter on the ice, glance at me, slowly walk away, linger again along the cattails, then walk all the way across the pond to the green water on top of the ice on the other side. Two tiny ducks—goldeneyes—have settled into a melt-spot in the ice, and with their typical loud whooshing sound a pair of mallards pops out of the dark sky, banks sharply, and on set wings descends gracefully to land near the geese. From all sides come the sounds of the red-winged blackbirds, grackles, and one of the first returned song sparrows. The wind picks up. More blackbirds fly over, singly and in small groups. A raven rises in the wind. The Sedge pair female examines sedge hummocks, sits down on them, and puts her wings to her sides, turning round and round, "trying it on" for a possible nest site.

And I continue the next day, March 31.

> There is an almost full moon as it is getting light at 5:30 A.M. The redwings and robins are singing under the clear sky. A glorious day. More ice has melted. The Sedge pair walk on the remaining ice all the way to the beaver lodge and check it out closely, then they walk all the way back to the other side of the pond to a patch of green

water on ice and paddle to the sedges, where she again sits down and "tries on" other potential nest sites. Three green-winged teal paddle next to the ice. Two are males in their nuptial plumage, that includes a beautiful iridescent green headband on their cinnamon heads. A mallard drake raises himself up out of the water exposing his light chest and beating his wings. A large beaver swims within five feet of both the geese and the pair of mallards, and none pay it any attention. Another beaver pops through thin ice, hauls itself out onto it, and perches there like a figurine for twenty minutes. A barred owl hoots repeatedly from the forest nearby.

As the sun comes up and illuminates the yellow-brown sedges along the black water on the other side by the white ice, a mallard pair swim up to the ice and hop on. The male's feet and legs shine bright orange, and the direct sunlight illuminates his emerald green head so that it seems to glow. His, and the teal's nearby, are the only green on the bog. It does not get any finer than this, I think, except possibly that any day now the Sedge pair, who have the first claim to the pond this year, may start their nest.

Early the next dawn I jump up out of bed with a jolt. There is a *huge* goose clamor down in the bog. I run down in a steady rain and through thick fog that hangs in the trees. There, on the pond next to the beaver lodge, are fourteen geese.

At first they seem like an amorphous mass within which there are occasional chases. But after watching them for two hours, I decipher details. There are two pairs and one individual who tend to stay separate from a main group of nine. The Sedge pair is one of these pairs, and it acts possessive by frequently perching on top of the beaver lodge and coming off it to swim around it every once in a while with much gaggling at the other geese who are all around. The other pair are . . . Jane and a mate! (It does not look like Pop from a distance.)

The odd, unmated goose leaves the group, swims across the pond, and approaches me, but then turns around and returns to the periphery of the crowd. I suspect that it could be Peep, but her "face" is not a distinctive one that is easy to identify from a distance.

The hubbub gradually dies down, and by midmorning some of the crowd of nine make signs to leave. Excitement mounts, and flock liftoff occurs. But the two pairs and the lone goose stay.

The five remaining geese apparently had much to settle and nothing among geese gets settled without a lot of noise. Their commotion lasted through the day and through the following night. Their cries were at times so strident that I wondered if another crowd had returned. I went down the driveway (April 1) to check at 5:30 A.M. but instead of finding five geese I found only two pairs. The lone goose (I suspected it was Peep) was gone. Curiously, most of the noise was now coming from just one bird—Jane's gander. I got a good look at him, and it wasn't Pop at all. It was a newcomer who, by continually following Jane and lowering his head and shaking it, gesticulating and calling loudly, acted as if he was strenuously trying to get her attention. For the hour and a half that I watched, he hardly ever let up. She, meanwhile, seemed almost nonchalant. If his courting succeeded, he would become Jane's third mate in three years. He might have an uphill battle—I suspected Jane was still waiting for Pop.

Surprisingly, the two pairs showed little mutual intolerance. When I first came in the morning, the Sedge pair were in the water next to the beaver lodge while Jane and her gander were within feet of them, directly on the lodge. Both pairs showed interest in the lodge and they occupied it alternately. Whenever the Sedge pair got on, they were always gently evicted by Jane and her new beau. There was no overt antagonism. The Sedge pair just calmly walked off the lodge into the water when Jane and her suitor came, and then returned onto it when they left. By the next day, however, Jane and her new mate started to show more active possessiveness when the Sedge pair trespassed onto the beaver lodge.

April 4, 2002.

Jane is again at the lodge, but seeing me pour grain onto the ground she paddles across the pond and comes to feed almost at my feet. Her new mate, whom I'll now call Harry, stands tall and hangs back. Within a few minutes after Jane comes to me, the Sedge pair take the opportunity to paddle over to occupy the lodge she and Harry have just left. Jane notices. She stops feeding, stares across the pond over to them, begins to call loudly, and then swims back out into the pond to her new mate. Harry has up to this point been unconcerned, but now at her urging he gets excited and shows his support by lowering his head down to the water and vigorously gesticulating with head-shakes and honks. As she takes the offensive and flies to the lodge to evict the Sedge pair, he takes off as well and flies beside her. A melee breaks out when they reach the trespassers on their chosen nest site. I can't make out the blow-by-blow details, but when it is all over Jane and Harry have prevailed. The Sedge pair leave, and the victors bathe vigorously at the edge of the beaver lodge, throwing water over their backs, rising up, and wing-beating, doing the equivalent of chest-thumping. Then they climb onto the lodge.

The retreating Sedgers call loudly and circle over the pond and the lodge to then descend to their old haunts at the southeast by the sedges. As they fly over and call, Jane and Harry answer just as loudly. It's settled! I feel confident that Jane now really has secured a new mate, Harry, and furthermore she will nest on the abandoned beaver lodge. Today they act like a couple. They've synchronized. She, the original pond owner for at least the last three years, may have lost Pop, her mate of last year, but she has regained her domain with the help of a new, vigorous mate. They own the prime nesting spot on the pond this year since Jane's old muskrat house has by now dissolved in decay.

On the same day that Jane and Harry secured both pond residency and ownership of their beaver lodge, I also saw Jane minutely inspecting a spot on the western side of the lodge, the precise place where she had nested three years ago.

Battles with other incoming geese continued, but the two pairs cooperated in evicting them. The next day at 6 A.M., I heard geese calling, and a pair came flying directly over the pond in tight formation. Their wing-beats did not slow down, and I presumed they would fly on. Both of the now-resident pairs on the pond seemed to know it, too, because they did not react.

Five minutes later a single noisy goose approached and this time both pairs made excited calls long before this goose set its wings to descend. The lone goose landed on the remaining patch of ice between the two pairs. Jane and Harry incited each other and flew off in unison for the attack. The goose escaped to the other end of the pond, and there faced the Sedge pair, who also gave chase. Meanwhile, Jane and Harry swam up along the shore all the way into sedge territory, to continue the attack. They swam within seven feet of the Sedge pair to pursue the single goose, and even in such close proximity the two pairs ignored each other now that they had a common enemy.

The battle against the single goose continued as one or the other pair sought out this very persistent bird, who ended up taking a lot of abuse before finally leaving a half hour later. Throughout all of that time she was chased from one end of the pond to the other and the two pairs routinely made intrusions on each other's pond space. They at no time chased each other and they continually showed solidarity with their respective mates. Pair members incited each other, took off together, attacked the stranger together, and afterward bathed together.

Jane and Harry returned to the beaver lodge after every foray against the lone goose, while the Sedge pair returned each time to the sedges at the east end of the pond. The persistent intruder had barely been repulsed when, less than fifteen minutes later, a group of six high-flying geese approached. They almost passed on, but then one individual near the end of the queue veered off to return. All the rest then about-faced also, set their wings, and descended onto the open water around the lodge. Jane and Harry attacked. The struggle continued for about a half hour. Curiously, this time the Sedge pair stayed silent and totally uninvolved, even when one of the six flew into their area of the pond. For the most part, Jane and Harry did not chase the six geese more than 10 to 15 meters at a time, but instead climbed to the top of their lodge as the others milled all around it. Relative to their spirited battles with the single goose, they showed little passion for conflict with the six newcomers. Those six left as a group after 35 minutes on what looked like their own volition.

Four days later, on a cold, rainy, early-April morning at my station at the edge of the pond, I again heard a (the?) lone goose coming from afar. The approaching goose instantly put Jane on alert and she called then as well, inciting her mate. Both she and Harry attacked as soon as the newcomer landed, and she was again chased from one end of the pond to the other by both pairs. Finally one of the Sedge pair managed to jump onto her and hold her under the water for a few seconds. It looked like rape. The attacked goose then flew to the other end of the pond, where Jane and Harry resumed the chase. Finally the much-maligned goose flew up, circled the pond a few times, and left. But in a minute or two she came back. The chases resumed. Finally acknowledging defeat, the goose left, again making long, drawn-out, sad-sounding calls that continued unabated far into the distance. I have to admit that in my reading of geese emotions from their calls, hers gave the impression of anguish. (I suspected this goose was Peep, because she acted as though she had a claim to the pond and because Jane's vehemence suggested that she recognized an old rival.)

Jane and Harry's coexistence with the Sedge pair was not to remain permanently without conflict. On April 12 there had been clamor at the pond most of the night, and at dawn I saw a full-scale quarrel in progress. Jane and Harry were in the Sedgers' territory, chasing them even there. Why were they now suddenly attacking this pair?

On the next day I knew. Harry was perching on top of the beaver lodge while Jane was busily making a nest mold next to him. She turned and squatted, pulling vegetation toward her with her bill and tucking it along her sides. A third pair of geese flew in and landed on the opposite end of the pond. Before, Harry had always been very vocal before and during any attack. Not today. Today he was all action, no talk. He flew off the lodge and headed straight across the pond directly at them. A twelve-pound goose flying at 30-plus miles per hour generates a lot of forward momentum. Attackees capitalize on that to wait until the last moment before jumping to the side rather than forward, so that the attacker rushes on by rather than into them. But today Harry went to the considerable trouble of beating his wings vigorously to bank sharply in flight and continue the attack even in the air. His attacks were fiercely unrelenting, and the pair made haste to leave the pond. But even then Harry remained in vigorous pursuit. When he came back, he landed next to Jane on the nest and after their mutual greeting he held his head high and scanned all around.

I later learned that, given a little more than one day to lay each of their eventual six eggs, and twenty-eight days of incubation, egg-laying must have begun on or near April 11. That is, his aggression coincided with the time when she would have had to mate (several days before and during egg-laying) to fertilize her eggs. All the legitimate matings (of pairs) that I saw occurred a day or two before or during the time of egg-laying. Selective pressure for the male to ensure his paternity should be particularly intense in species where the males make a long-term commitment to one mate. Jealous mate-guarding is a prime mechanism that helps guard investment in a long-term mate.

After Jane began incubating, the early-arriving (but ultimately late-nesting) Sedge pair again took on an active role in bog defense, while Harry's role as a territorial defender declined but was not extinguished. There were clues here that would eventually help me to understand increasingly strange goose behaviors.

13 JANE AND HARRY

I wanted to see if Jane and Harry had started to lay eggs so that I could determine the hatching date (after twenty-eight days of incubation). I also wondered if Harry would respond to my kayak by hiding behind the nearest clump of sedges, as Pop had done.

The pair are foraging along the north shore when I launch out into the pond on the evening of April 15. They are dipping with tails up and heads down to reach food at the pond bottom. I know there is a lot of potential food down there. Six sweeps with my insect net in the water in the dead vegetation near shore have yielded two minnows, three newts, two mayfly larvae, one damselfly, and one dragonfly nymph and also a backswimmer, a notonectid water bug, and two small hydrophilus water beetles. I have seen little before my blind sweeps through the murky water and I suspect the geese feed blindly, too. Harry often treads the pond bottom near shore with his feet, sending up clouds of mud and bubbles, and he then reaches down with his long neck into the stirred-up soup. Jane continues her dipping as I start to paddle across the pond. Harry looks up, but then resumes his foraging; to him the kayak is of no immediate concern.

I reach the beaver lodge. Yet no nest is visible even though I have distinctly seen Jane making a nest mold. Has she buried her eggs? I gently

Harry (top)
Jane (bottom)

poke in the rotting vegetation of the ancient beaver mound with a short stick. I strike eggs. I can now start the countdown to hatching, perhaps after a day or two more when incubation will begin.

Incubation starts only after all but the last egg has been laid, to ensure that all the eggs hatch within a day of each other and so that no gosling gets left behind. Jane has not yet left any down feathers in the nest. Such feathers would reveal the nest to sharp-eyed and perhaps even to not-so-sharp-eyed nest predators, and it takes nine days to lay a full clutch of six (one and a half days per egg). During that time the eggs must be hidden because the goose is not on or generally near them to hide and protect them. Additionally, she should not shed her insulating down feathers from her belly so long as she still daily swims for hours in the icy water to forage. However, after incubation begins, those feathers are more useful off than on her as a nest lining. Belly feathers at that point will only reduce or eliminate the skin contact with the eggs that is then required.

Harry and Jane stay away from the nest and paddle conspicuously in front of me out in the pond. However, as soon as I leave they leisurely paddle back to the lodge (nest) and swim around it. A beaver comes as well. Harry jumps at the swimming beaver and it dives to resurface fifty feet away.

Later in the day both geese come to me and Jane feeds on grain. During the last month a pair of mallards, who have seen my daily feeding routine with the geese, now also approach as soon as they see me coming. Jane ignores the drake but bites the hen and pulls out a billful of her feathers.

Today Harry decides when it is time for Jane to leave the food and return to the nest. He makes soft grunts and wags his head slightly. That is enough to get her moving out onto the water, where he waits for her. They then both paddle across the pond back to the nest on the lodge. Jane climbs onto it and preens. After slumping down for a few minutes, she rises again and reaches to the up-slope portion of the beaver lodge to pick up soil and debris, and in quick motions of pick-and-throw she dumps one billful after another next to her for the nest rim. While she is thus occupied in nest construction, a wood duck drake swims up to the other side of the lodge to perch there and preen. An hour later Jane is still busy rearranging the beaver lodge for her nest. Finally she reclines, possibly to lay another egg, and tucks the loosened material all around herself. Harry stands within two feet of her the whole time and surveys the bog, but once she starts incubation he will stay clear of the nest.

Meanwhile the Sedge pair goose is testing still more sedge hummocks as potential nest sites. So far none seems to be just right. Given this pair's already very long delay and their indecision about where to nest, I suspect that they are a young couple (possibly two-year-olds) who are only play-nesting.

I awoke in the predawn the next day hearing approaching geese, and when I came down I saw a third pair on the pond being engaged by the Sedge pair. Harry and Jane left their lodge and came to join the fight, and when they did so the female of the Sedge pair tried to hide; she put her head down with her neck horizontal to the water and swam behind a screen of beaten-down cattails. There she hid by lying flat on the water, her neck level with it, and she did not move. The intruding pair did not leave until an hour later, but their departure was ten minutes after their last fight with Harry and Jane, not during it. It did not look as if they left because they were beaten. I wondered why the Sedge pair female first fought, and then after being helped, hid.

I heard a new redwing sound today, a series of sharp, abrupt *tit tit tit* . . . followed by throaty *churr churr churr.* The females had finally arrived. Would they now be welcomed by the strongest males? I saw no males "defending their territories" as they are supposed to according to the dogma. Instead the four chases by redwings that I observed were by groups of males in very rapid flight behind females. In one instance six males were zipping behind one female. If the males were "displaying to set up territories," should they not have been staying in specific areas, paying some attention to each other, and letting the females come to them, rather than chasing them off? Instead, they seemed to have it backward. They were tolerant of one another. I had identified one male by his voice; he and another male regularly perched within two feet of each other, and they ignored each other day in and day out. Yet they sometimes ranged widely over the marsh, and they appeared to chase off some females while tolerating others. Of course, appearances are often deceiving; that's what makes close observations fun, and necessary.

The common grackles, *Quiscalus quiscalus,* are a picture of opposites compared to the redwings. Males and females stay peaceably together. Today, April 16, two pairs of them spent much time at the cattail point jutting out into the pond, where several had nested in past years. I suspect they'll soon nest there again in a small cooperative group.

Most exciting to me has been the variety of ducks that stop at this little beaver pond on their spring wanderings. Several species arrived as soon as the first open patch of water appeared. Aside from the resident mallards, wood ducks, and black ducks, I've seen buffleheads, common and American mergansers, ring-billed ducks, green-winged and blue-winged teal, scaup, one eared grebe, and one pied-billed grebe. At dawn I now see regularly as many as three pairs of the brilliant white common mergansers with their red bills, accompanied by their russet-headed females. Soon after morning light, one by one the merganser pairs lift off, circle the pond, and leave. All are gone before the sun shines. By evening they—or others—are back. Some of the other ducks stay only for a day on this migratory stopping place. The mallards and wood ducks are regulars. Today, with record temperatures of 90°F in the shade, marks the end of the visits by the beautiful and varied ducks and also the beginning of the spring transformation. In one day the hillsides leading down to the water have flushed green, pink, yellow, and white from bursting leaf and flower buds of maples and shadbush.

April 20, 2002.

At dawn a wrought-up bevy of seven redwing males noisily converges from all of their neighboring "territories" in the cattails for a by-now-usual morning round of cooperative grackle-thrashing. Grackles are nest-robbers. Presumably one was just caught in the act.

I don't know whether it is before, during, or after this hubbub, but around 7 A.M., I hear a loud roaring noise. A huge jet plane taking off or landing at the Burlington airport? But that is over ten miles distant! A new super-large 74900 or some such new-issue model that has never landed there before? The roaring continues and gets even louder. Engine trouble? I start to feel vibrations through my feet. A *very* big aircraft, I think.

I am totally wrong. We filter our perceptions and match them to our experiences and expectations. (I later learn that our area has been rocked by an earthquake that measured 5.1 on the Richter scale.)

May 1, 2002.

Jane had been steadily incubating for almost a month and I expect to see the goslings at any time. The Sedge pair dawdled in their nesting attempt, and after rejecting innumerable sedge hummocks, they eventually chose the platform I built two years ago as a nest site for Peep and Pop. The Sedgers laid only three eggs, and the goose was only mildly attentive in incubation. At first the goose spent hours off the nest, then days. (All of the experienced nesters laid five to six eggs and either took only rare and brief breaks or incubated continuously.)

For the past month, Jane has been continuously on her nest. I never see her off, and Harry stays away. He has only recently started to attend her again.

Today he is at her side, and at 6:23 A.M. he makes gestures that indicate he wants her to leave and follow him. That seems odd: Why would he want her to interrupt her incubation? Just as curiously, she responds to his summons by rising from the nest, and both fly off! I have seen Pop and Peep behave similarly, but I am still flabbergasted to see them both leave the pond entirely. Since nesting began I have not seen either leave it.

They start to fly toward the west over the beaver dam, but she seems to reconsider following Harry and drops down onto the water flowage immediately on the other side. Within a couple of minutes she returns to the nest. He also returns, but he then again gesticulates for her to leave. Again they fly off in the same, westerly, direction, and this time I hear their voices trailing off into the distance.

Three things seemed odd about this behavior. First, *he* initiated and incited her to leave the nest, which I have never seen him do before. Second, he wanted her to accompany him on a long trip away from the nest. And third, the excursion was apparently not to enable her to feed, since there was food nearby and it was now more plentiful than ever. The sedge plants were now in full bloom all around the pond, and these flowers are one of their favorite foods.

In science, it is hardly possible to make significant observations, or even to notice much, without at least tentative expectations or hypotheses. Anomalous observations, provided they are detailed enough, are valuable (and exciting) because they provide clues to understanding the larger picture. Patterns continue to emerge gradually from an accumulation of details until the significant patterns stand out from the insignificant ones, finally providing the links to a viable theory. Patterns are discovered, and as a scientist my task is to investigate them to show how these patterns make sense of a piece of nature, in this case, the behavior of these geese that sometimes seems so idiosyncratic. The observations I have just made don't yet fit into any consistent theory, and that is precisely why they are important. They mean that something is up, something I don't understand but that could provide a connecting clue to a larger picture.

Harry and Jane returned from their mystery excursion from the nest after 33 minutes, which is an unusually long time for a goose to leave her nest. They had flown west, in the direction of the Sunset Lane pond. Did Harry want to scout out territory (possibly in the area where he grew up) with Jane, to find a place where they might lead the soon-to-hatch goslings?

May 13, 2002.

I am surprised and cheered on this drizzly and cool morning by three big, brown birds circling the bog. They are the first bitterns I have seen here. Two fly to the upper end by the sedges, while the third drops down by the beaver pond beyond the lodge where Jane is incubating, with Harry again close by her side. At first the bittern looks like a brown skinny stump but a few minutes later the "stump" melts down into the sedge and disappears.

The Sedge pair is feeding on sedge flowers in a cove far from their now almost abandoned nest. An hour later both drift by their nest without paying any obvious attention to it. Instead, she bathes, then hops onto the unoccupied muskrat lodge where Jane nested last year, pulls up a leg, and tucks her head into her shoulders to sleep. Her gander paddles around a peninsula of cattails. Occasionally she raises her head and looks in his direction. She calls and finally she jumps off, swims to her nest, and settles down to incubate. However, owing to her gross negligence so far the eggs are surely long dead. (After this she sat on the nest a few more times, but only for short durations before abandoning it totally.)

In the next two days, May 14 and 15, Harry got ever more attentive to Jane at the nest. I checked numerous times each day, and he was perched close to her each time. At 4:30 P.M. on May 15 she held her head down, rather strangely I thought, and I immediately suspected that the goslings were hatching. Given my strong hunch, I paddled out to check.

Harry (unlike Pop) was again unalarmed, if not totally unconcerned, by the red kayak. Whereas Pop, a much tamer gander, had hidden from me when I went to the pond and even when I looked toward him, Harry unhesitatingly approached to within several feet of me and the kayak at the nest as I gently prodded Jane with my paddle to induce her to stand up. So Harry's previous nonchalance with the kayak was not a function of the nesting cycle as I first had thought.

And I saw what I had expected—goslings. There were several already fluffy, dry ones and one that was still wet. (She was not willing to stand up long enough for me to make a count, and I was unwilling to disturb her further.)

It was a windy day, and as I was trying to steady the kayak and paddle around the nest, the gander of the Sedge pair swam out toward us. He held his head high and seemed curious. Harry ignored him.

I hoped to witness the goslings leaving their nest, and about twenty hours after the last hatchling came out of the eggshell and was dry, I again parked with the kayak next to the nest and waited. As before, Harry remained indifferent to the kayak and stayed next to me. I was glad because it meant that, unlike Pop, Harry would not delay the natural time of nest-leaving by Jane and her young.

Like last year, I was captivated by the endearing scenes of the goslings tumbling under and around their mother. Whenever she stood up they got up, too, pecking at everything in sight, shaking, preening, flapping their wing-stubs, scratching themselves, and then all going back to sleep under Jane when she sat down. Harry spent all of his time about two feet from them, standing calmly on the top of the beaver lodge, while my kayak was parked against it.

Harry waited until a little past noon (on May 16) before signaling that it was time for departure. I could read it as plainly as written text. It began as he calmly walked off the lodge, entered the water, then swam directly in front of Jane on the nest and made the head-wagging "Let's go" signal to her. Then he remounted the lodge, continuing to signal. As with Pop's head signals after incubation began last year, his were unaccompanied by honking. Nevertheless, Jane responded by rising up in the nest, fluffing out, shaking herself, then standing tall and performing a series of vigorous wing-flaps. The six youngsters all stood up with her, and some of them reached up to touch her bill as she put her head down to them. She made soft little grunting calls and slowly eased off the nest and slipped into the water. One of the goslings ran ahead of the others to the water's edge and without a moment's hesitation jumped in and started paddling as though it had always done so. The rest of the brood followed and plopped in one after another. Harry stationed himself next to Jane, but the goslings gathered in a clump, and only around the mother's head. Within seconds she started paddling toward shore and the tight little knot of her six goslings broke out into a line behind her. Harry brought up the rear, and I in turn followed respectfully in the kayak and continued to take baby pictures. She started paddling faster then, but only as fast as would allow her young to easily keep up. As soon as she got to the sedges near shore, she started ripping into them in a feeding frenzy. She wasn't choosy about looking for tender parts or selectively eating flowers. It was perhaps her first meal in a month; I had not seen her eat during the entire month although she could have fed at night or when I wasn't there. (In contrast, the Sedge pair goose had fed often and long, while Peep had fed daily but only briefly and hurriedly.)

The whole family traveled along shore toward the big beaver dam over which the parents had flown five days previously. Harry took the lead there, stopping frequently with head high, to look over the dam. Then Jane and the goslings swam back and forth along the dam as though looking for a place to cross. After about fifteen minutes Harry went over the dam and out of sight into the bushes on the other side. Jane and the goslings still hesitated. She stood for a minute or two looking over the dam into the bushes, and then she and the six goslings also left the pond. I could not follow them through the dense *Viburnum* thickets without making a huge commotion, and if I had tried to do so they would have been alarmed because they would perceive me as chasing them. So I had no choice but to leave them alone. However, I hoped to see where they ended up. They might now traverse territory on foot that they had only seen from the air, although Jane had presumably explored this route before. But maybe Harry would continue to lead the way.

14 GETTING TOGETHER

It seemed odd that every goose family vanished from my beaver bog within two days after the young hatched, although for a long time I had no problem coming up with rationalizations that canceled the need for thought: They hid, or coyotes, snapping turtles, or raccoons got them. The problem is that, until last year, there was no data to back up any of these speculations, and I had not seriously considered a long shot: emigration to another pond. As already mentioned, last year the breeding attempt of the pair at Sunset Lane was aborted, but two pairs of geese with goslings came to rear their young at this pond. One of these pairs was Jane and Pop with five young, whose brood suddenly doubled to ten young! Why do the residents stay at that particular pond and why do other geese come there from other ponds? Why were some young adopted and others given up? Would similar behaviors be repeated or were they anomalous?

It seemed preposterous that goose parents would lead their just-hatched goslings on a dangerous journey from one pond they had just defended fiercely to another one occupied by other geese that was miles away, where all would then get along in peace. Nevertheless, that's what they did. Undoubtedly overland journeys through the woods and fields are risky and they would not be undertaken except for a large benefit. It was necessary to find out who was going where so as to figure out why.

Last year, Pop and Jane left our pond, taking their family to the Sunset Lane pond, where they joined the resident family and their five goslings. I wondered if Jane would again arrive at the Sunset Lane pond in 2002, this time with her (six) young and a new gander. I was skeptical, because her new mate, Harry, might not know whether or not to leave the beaver pond, nor where to lead his family if indeed he did the leading. Maybe Jane rather than the gander would undertake to lead them through the woods and over the mountain.

I did not see Jane and Harry arriving at Sunset (although I checked twice daily), and it would have been highly fortuitous if I had, given that they could come at any time of the day or night over the course of a week. But geese with young had already arrived earlier: A pair had nested there this year and produced young and that family of five goslings had increased "overnight" to nine, presumably because another pair had come with four young and then left without their goslings. Following that, a second pair had come to the pond with four goslings, stayed, and then ended up with ten goslings. Were those six extra ones Jane and Harry's? That is, two pairs (one possibly being Jane and Harry) had each brought their goslings and then departed, leaving them with baby-sitters. Two of the baby-sitters were the pair that had, this year, nested at the pond. However, these were no mere baby-sitters—they became foster parents. As we will see shortly, the biological parents, though being and remaining whole and healthy, left their young in others' care all summer until they grew to adulthood.

Goose families stay together for a year or more. The repeat pattern of adoptions this year was as much a surprise to me as the geese's interpond travel. Apparently, it had not been a fluke the first time I'd noticed it. There was no way of knowing if Harry and Jane's young had been dropped off at Sunset Lane, but it made no practical difference; the fact was that in 2002 at least two pairs had dropped off their young there. Furthermore, the two pairs of geese that remained at the pond associated as apparent friends, and together made up a superfamily of 23 birds with 19 goslings in tow.

Throughout the summer the two baby-sitting pairs foraged within about 10 to 30 feet of each other, although the young, whether offspring or adopted, stayed strictly with their parents within the superfamily. All the geese at times intermingled, but when the two pairs of parents separated, then the young without exception remained in their respective groups of nine, or ten, and followed their parents. There were no antagonistic interactions between these two pairs although there could have been some squabbling on first arrival.

The possibility that something happened to the parents so that they were unable to take care of their young was put to rest on June 9. Our beaver pond had been devoid of geese for about three weeks (since Harry and Jane had left with their young). There had been no honking of geese at the bog, and those geese with young at Sunset Lane pond had also become mute. But on this morning there was loud, excited honking at the beaver pond. I immediately rushed down to check on the hubbub because geese honking in the summer is out of the ordinary.

It was Harry and Jane. They were without young and they were paddling rapidly all over the pond. They were emotionally tense. Both were calling at steady and frequent intervals. I yelled out to them, and Jane stopped, looked at me, and started paddling over to me. But after advancing less than fifty feet, she turned back to her mate. I called again, more insistently and continuously, and this time she swam to within about twenty feet of me, and I could positively identify her by the small nicks and pips on her white face pattern. Harry had started to follow her, but then stayed back after climbing onto the nearby muskrat lodge. Using my binoculars I then positively identified him as well.

They had not come back to "their" pond and bog in order to feed. I was almost startled to see Jane come to me, because I had nothing that she could want, except friendly familiarity. I brought them grain, which they refused, and within only a half hour they flew off and I did not see them again that summer and fall.

Presumably they went north where the other nonbreeders of the "resident" Canada goose population migrate in the summer. Bill Crenshaw from the Vermont Fish and Wildlife Department found by radio-tracking that a goose breeding in Vermont, whose nest was destroyed, then migrated north to the Quebec molting area where the birds lose all their flight feathers, to join the other nonbreeders (personal communication). Perhaps Jane and Harry had checked in on their familiar haunts once more before departing—before severing their ties to the place. I would give a fortune for the ability to experience *what* they felt. Only one thing was certain. They were charged with heightened emotions.

In the summer of 2002 there was one more major event at the Sunset Lane pond, and I happened to be there to witness it. When I had driven by the pond in the morning on June 12, I had, as usual, stopped and seen only the megafamily of twenty-three geese. At noon, when I came by again, the two pairs were grazing near each other and about fifty feet from shore. A group of three adults who had arrived on June 8 were grazing on the opposite side of the pond. I stopped to determine if geese with young would have the predictable response of holding their heads up and scanning more than those without. (The three adults without young grazed on the grassy hillside almost uninterruptedly, each scanning only 5 percent of the time. In the group with young, however, there was always at least one bird scanning, sometimes even three. On a per-bird basis, the parents scanned over eight and a half more times than the other adults. Parenting thus has at least one cost. It compromises feeding time.) While watching the geese for a by-now-scientifically-trivial reason, I had the serendipitous luck to witness something relevant to the story of these geese—the arrival at the pond of a group composed of two families that would from then on stay the rest of the summer and fall.

One pair of the two newcomer families had only a single gosling, and the other pair had four. Neither of these two new pairs who had aggregated into one group was familiar to me; I had not seen or recognized their face patterns before. When this two-family group of geese approached the pond on foot, my attention eagerly shifted from watching scanning versus feeding behavior to looking for reactions from the residents. These were not long in coming.

The gander of the resident pair with the nine already gray "teenage" young started to amble toward the two new pairs that had just started to graze. Suddenly he lowered his neck and angling his head up he ran toward them. His goose joined in the threat but not in the attack. The geese did not honk even once, unlike all their aggressive interactions until several days prior to egg-hatching when they honked loudly and continuously. I presumed, however, that they were hissing, since their bills were open. The nine young of the attacking pair all stayed back, bunching up into one tight group, and they raised their heads high and watched. Within seconds the resident gander was in a pitched battle with the newly arrived gander of the pair with the four young. The participants were silent, but the "cheering section" of mates and goslings peeped loudly throughout the fight, that lasted several minutes. Both ganders locked together in close combat. They moved methodically but silently, biting, hitting, thrashing, and plucking each other. Clumps of feathers flew off in all directions. It was now the time that the geese with young (those who did not migrate to their northern molting areas) were shedding their flight feathers, and the fight greatly accelerated the molting process. (I later found twenty-two large wing feathers and dozens of small feathers at the scene of battle.)

After the fight, the resident gander ran back to his mate and nine young. They greeted each other with head-wagging and he then raised his head high and, walking slowly and deliberately, led them away.

I did not see another fight that summer. The two newly arrived families with their total of five young usually stayed as one unit, while the other two pairs with their total of nineteen young remained in a separate unit. These families ignored the threesome of adults without young (who left within a week), even when they were within about thirty feet of them. Like other nonbreeders, the three probably left in June to migrate north to their molting area in the upper Ungava Peninsula of Quebec where they would spend their month of vulnerable flightlessness.

I kept a lookout at our pond throughout the summer as I waited for Peep to come back. I saw no sign of her, but observed the geese at the Sunset pond daily. All of the young who had arrived at that pond had avoided the threat of predators, although there was a fox den nearby. By early August, all 24 young were still alive and in their respective family groups. They were then grown and difficult to distinguish visually from the 8 adults that still accompanied them. However, I could easily differentiate the young from the adults at close range by their soft but high-pitched peeping calls. At times the groups of 23 and 9 geese came together and looked and acted like one anonymous flock of 32. With the Sunset Lane pond population increasing from 12, to 18, to 32 in three consecutive years, it was tempting to extrapolate that there would be an even larger population in 2003

15 THE SEDGE PAIR

My foray into the life of the bog and its inhabitants had become an adventure. It had also turned into a pilot study that could potentially lead to a full-blown scientific investigation. However, hoisting the banner for that route would involve catching the geese, marking them with plastic wing tags or cumbersome neck rings and metal leg bands, strapping radio-tracking collars onto their backs, rounding up the goslings and ringing them, videotaping the nests for long-term surveillance, dropping the rest of my life to beg for money to buy equipment and technical aid, and possibly recruiting students to do the work and have the fun. Pursuing this trajectory would do more to advance one's reputation than make discoveries. And so by spring of 2003, I felt I had reached a point of diminishing returns regarding benefits, and was therefore starting to ease up on the goose story. Instead, I spent more time watching the blackbirds, who were not living up to their reputation for polygyny. If I hadn't known better, from what I'd seen so far I would have concluded that their males team up to evict females from their territories. I concentrated on watching them from March 21, the day they arrived, until mid-May when they started to nest. Meanwhile, I continued to keep one eye on the geese.

March 21, 2003:

Six redwing males have arrived, even as the bog is still under deep snow, and the pond ice is almost two feet thick (I chopped through it, all the way to the mud on the pond bottom). The six spread out over the bog as if they know their places, rather than fight to gradually establish exclusive territories. Their *oog-la-eee*'s reverberate from one end of the bog to the other. It looks as if boundaries have been already established. But just when I think all is settled, I see a bird leave his "territory" in the bog and fly high into the air and straight to the west clear across the pond some 100 meters distant. I follow him with my eyes, curious why he would "desert his territory." Then I see him engage in an aerial battle with another male. The chase continues for fifteen seconds, and then both birds return to the bog, circle over it, and settle 30 meters apart on the tops of trees at the edge of the bog

A flock of six grackles arrive near noon. They circle the bog several times—unlike at any other time—then they land in the trees. By 4 P.M., they fly down into the cattails where they have nested in previous years (and will again later this year). I have caught their arrival back to their "home" ground.

March 24, 2003.

I'm down by 6 a.m. Several redwings are "on station" within the bog, but my attention is almost immediately riveted on two that are on the bushes directly at the edge of the bog. They display to each other, taking turns making the *oog-la-eee* calls with the simultaneous shoulder shrugs and displaying their bright epaulets. I end up watching these two uninterruptedly for ninety minutes, during which time they twice tussle in knock-down, drag-out fights. (The victor stayed all spring and summer on his specific patch of cattails where he had only one female nesting. Another male, his neighbor, visited often, and these two males were commonly two to six feet apart. They sometimes cooperated in chasing other males. I called this bird Fugel and recognized him by his distinctive voice. He sang a two- or four-syllable song—*witch-witch* and *oo-ee-chee-eer* unlike the standard three-syllable *oog-la-eee*.)

The Sedge pair.
Female top, male bottom

As in the previous year, the Sedge pair came back first (on March 28) and as before they immediately showed an interest in the muskrat lodge where Jane had nested before and from which she and Harry had evicted them. However, the pond ice was still thick, and it was still too early for geese to start nesting. They were, nevertheless, already making their claim on the pond. Finally, on April 16, they started to lay eggs and they produced a full clutch of six on that previously contested lodge. Until then Jane and Harry's absence was conspicuous, and no geese had come to pose a challenge to the Sedge pair. Finally two geese started coming daily and nightly, but by that time the Sedge pair had already finished laying their eggs and were at the height of their aggressiveness. The latecomers didn't stand a chance. Usually they came at dawn and again at dusk, and each time the Sedge gander went at them in a fury driving them out in a minute or less.

The Sedge pair soon ruled the bog. As I later confirmed, the "intruding" pair were Jane and Harry. Of course, Jane could not hold on forever and wasn't expected to, but for her to be summarily ousted by these two young upstarts whom they had tolerated (and who might have been her offspring) still took me by surprise.

Throughout the whole period of incubation the Sedge gander remained totally mute and almost always hidden, and I only twice saw his goose off the nest in the thirty-one days she was on it; she had reformed her wayward and youthful ways of last year when she'd been lackadaisical about incubation and abandoned her three eggs.

Although the geese were sedentary and quiet for the month, I was entertained not only by the grackles and the redwings, but also by the male snipe who did his best to be noticed. High above the bog he circled and I could sometimes follow his uninterrupted sky dance for fifteen minutes at a time. It was an awesome athletic display, accompanied by a throbbing flutelike whistling sound that you could hear from predawn on into the morning. He was the sky flute. He fluttered quietly upward and after reaching the apogee where he was almost out of human sight, he still beat his wings just as rapidly, but then accelerated on a downward plunge. As his downward momentum picked up, his tail feathers spread laterally (to engage the outermost feather to the wind) and they vibrated like a flute's reed to cause his sky-song; but you heard a pulsing-throbbing sound that was in synchrony with his wing-beats (because of air-pressure changes affecting the sound). The sound picked up steadily in pitch with speed and reminded me of a plane on a suicidal dive. Then the feathered meteor pulled up, retracted his tail, silenced himself, and again repeated the roller-coaster ride as he circled the bog over and over again. Eventually he plummeted down with folded wings to begin his vocalization on the ground, a metronomic *tuck-a, tuck-a* repeated at frequent intervals. I feel a liberating satisfaction in entering the wild through such another's strange and wonderful orbit of existence.

This year the goslings should have appeared after twenty-eight days of incubation, on May 19, 2003. But I saw no goslings peeking out from under the female Sedge goose's wings.

Finally, when she had still not left the nest on May 22, I paddled out to check. She rose up and hissed (her gander stayed in hiding), and I saw only one gosling. The remaining eggs seemed cracked and looked crushed. Might they hatch by morning?

By the next morning she is perched about six meters away from the nest on another muskrat lodge, while he is lying down near her. He looks dopey. His eyes are not wide open and alert. He closes them frequently, as though he is depressed or has had a long night. Meanwhile, she rises up only briefly, and I still see only *one* gosling, who nudges her and tries to get back under her as soon as she sits down. Like Peep with only one hatching gosling, the Sedge female had also stayed three "extra" days on the nest. I now know that the remaining eggs did not hatch, and I must investigate the nest and at the same time see if she will defend it, as Jane did.

She slides off the nest as I launch my kayak, and leaves with her little gosling that follows as if attached to her. The gander brings up the rear in usual goose fashion. As I pursue the family, she shows not the slightest tendency to come back to try to seek the "safety of the nest." The family of three travels fast and far and disappears into the sedges on the other side of the pond.

I now turn my attention to the nest and see a sorry sight. Five smashed eggs are trampled into the bottom of the nest. The whole mess stinks, and I try not to inhale too deeply as I open the putrid eggs and see the deceased fully formed goslings inside. They have walnut-sized yolk sacs attached to their umbilical cords on their bellies, so they have died within days of hatching. They did not die from a chemical imbalance and/or a developmental defect, because they succumbed virtually synchronously after growing normally to just short of hatching. They are also not victims of a predator, because there is not one tooth or fang mark on them. I conclude that they have been victims of another nest attack, by a pair of nonbreeders of the same kind that Peep's nest experienced. Peep's nest was shallow and the eggs rolled out, but this nest cup is deep. (I had excavated a hollow in the top of the muskrat house in anticipation of it being used by the geese.)

Given the approximate age of the about-to-hatch goslings and their state of decay, I deduce that the nest attack occurred in about the second week of May. And I recall the curious behavior of the Sedge gander at that time. Referring now to my *notes*, I read that there was a huge commotion on the evening of May 10. The Sedge gander usually routed Jane and Harry within one minute, but this time the commotion lasted longer. Usually the Sedge goose stayed on her nest, and after chasing the trespassers her gander resumed his vigil in hiding, far from the nest. I seldom saw any sign of him. *This* time when I ran down to the bog to see what was happening I found her off the nest and he was accompanying her on her way back to it. As she hopped back on, he suddenly started vigorous wing-beats and at the same time sent up a spray of water by rapidly stomping on the water with his feet. His chest was forward and his neck arched high but his head was sharply bent down. I heard loud splashing as his feet beat the water. I had never seen anything like this before. However, I saw nothing amiss. She preened and settled down to incubate while he walked up onto the nest, stood beside her, and with his head up high jerkily looked all around for several minutes. Meanwhile, Jane and Harry continued to call loudly from the neighboring pond. I suspected they had just been in a fight. (But I then had no inkling that it may have been associated with a raid on the Sedge pair's nest.)

Why did Jane and Harry now keep intruding on the Sedgers' nesting? They had missed their own chance to breed. It was too late for them to start a nest on their own. There were no eggs to fertilize in forced copulation. (It was long after a male could inseminate another female.) It was also long after a female could hope to insert an egg that would produce a gosling into another's nest. Any eggs that she might add now would not hatch since the geese would leave all unhatched eggs behind. Were they spiteful because the Sedge pair had prevented them from nesting?

The day after I discovered the Sedgers' five destroyed goslings (May 24) I felt psychologically "down." I had often been unable to identify the frequent intruders. I didn't know why the pairs fought and why the eggs and young were destroyed, and I felt a loss at seeing the five dead goslings in the nest. Furthermore, I thought the goose show was over for the year. However, at dawn on hearing geese call in the bog, I jumped up instantly to be there, this time even *before* a cup of coffee.

All the noise was coming from Jane and Harry at the far western end of the pond next to the big beaver dam. I remained hidden in the now-leafed-out woods, and soon saw them cross the pond. They swam steadily and without swerving even slightly, went directly to the Sedge pair's just-vacated nest. (I had removed the dead eggs/fetuses the previous day to freeze them and possibly examine them later.) Nearing the nest, Jane lowered her head horizontal to the water to greet Harry. Then both hopped onto the nest and gesticulated even more to each other. They alternately shook themselves, but violently, like a dog shakes itself after coming out of water. Next they preened. They were charged with emotion, perhaps because they were now in full control of the nest.

I then noticed the Sedge pair, the owners of the vacated nest. They were watching from the south shore. After about ten minutes, the Sedge female started to swim over to Jane and Harry. Her one gosling and mate followed. They came close to the intruders on "their" nest mound, and then the Sedge gander rushed forward and attacked, displacing the usurpers who were now forced to park about twenty-five feet away. He, his mate, and their gosling then mounted the nest. He did his wing-beating ceremony, and both parents then did the same vigorous body shakes, followed by preening, that the usurpers had done before. The two pairs faced each other and kept their heads high.

A half hour later they were still facing each other, and then Jane lowered her head and advanced toward the Sedge pair and their gosling. Harry followed. As she got up to the nest, the Sedge female defended; and immediately the ganders joined the fray and a melee ensued. The Sedge pair's little gosling became separated and peeped piteously but it was soon reunited with its mother. The fight was over in seconds, and the Sedge family remounted their nest. Jane then left for the eastern side of the pond. Harry stayed and started to preen. Curiously, *he* was totally ignored by the Sedge pair, and vice versa. As in previous recent encounters between these geese, it looked as though the quarrels at the nest were primarily between the females.

At 6:35 A.M., 22 minutes later, the Sedge pair again departed from their nest. She led, as usual. Baby and mate followed. They started to forage, and now Jane and Harry advanced again and again mounted the just-vacated nest, only to be once more chased off by the Sedge pair. The contest over the now-useless nest seemed settled, so I left.

Two hours later, when I checked again, the Sedge family was nowhere in sight. Unchallenged, Jane and Harry had once again come back to reoccupy the Sedge pair's empty nest, again mounting it and inspecting it closely by looking into it and poking it with their bills.

Might Jane and Harry remember me? After they had swum off to the southeast side of the pond, I came out of the woods and called to them loudly from shore. Both looked, and then started paddling toward me. They came straight across the huge pond, stopped thirty feet from me, and looked me over. After a few minutes they started to wander back. I ran up to the house and came back with corn (which I had not used this year). Again I called them while I threw it out into the water. Again they came to me on the double. But they showed no interest in the grain.

Near noon, when I came down once more, the Sedge pair was still missing, and Jane and Harry were on the beaver lodge where they had nested last year. This time *they* called as soon as they saw me, jumped off, and came clear across the pond to me. We exchanged long glances, and then they slowly drifted away while I went back to the house. Two and a half hours later I heard their haunting cries as they were leaving, flying away from the bog, their previous home that had this year been denied them.

I do not know what motivated Jane and Harry to come to me. The pull of memory can create a nostalgia for the past. Perhaps they had recognized my power to provide safety and security, or had they been motivated by curiosity? Whatever it was that drew them to me, it went beyond the simplest and most conventional hypothesis—that they came for food. What drew Jane and Harry to me now, this day at this time, had deeper roots.

Jane, a long-standing resident at this pond, and her gander, Harry, had been denied breeding privileges by the Sedge pair. But like the nonbreeding pair at the pond the year before (who had repeatedly attacked Peep and Pop's nest) these nonbreeders had spoiled the newcomers' nesting attempt. Now they had ascertained that their rivals' nest was indeed empty, and I predicted they would not likely show up again this summer but would come next spring.

There were no geese in the bog the next day, or the next. I started to wonder about where the Sedge pair might have gone. They did not show up at the Sunset Lane pond. Indeed, contrary to my extrapolation from the previous years, Sunset Lane pond had not hosted *any nesting* geese this year. The pond was occupied only until the end of May by a new nonnesting couple who aggressively chased out other pairs. Finally, after they left, two pairs with

Sunset Lane pond with a superfamily of sixteen geese, three pairs with a total of ten young. Early June 2003.

4 and 5 young, respectively, arrived. They were later joined by a third pair with 1 gosling, and these 16 geese stayed together all summer long as one family that successfully raised their young to adulthood by the end of August.

Our beaver bog remained silent during the summer. That was expected. There had never been geese on it in the summer once the young had left. A day or two after the families with young left, the geese with whom the breeders had fought also stopped coming. There were, however, rare exceptions. Jane and Harry had come last year for a very brief visit after they had lost their six young (possibly to adoption by a pair at Sunset Lane). The same pattern would repeat now.

In the evening of the second day after the Sedge pair has left, I am surprised by a shattering of the stillness that I expected to last for the rest of the summer. I rush down to the bog, as I always do at unusual events, and I see the Sedge gander perched on the beaver lodge, with his goose swimming across the pond. Both geese keep calling uninterruptedly. Their calls echo from shore to leafed-out shore each at 25 to 30 times per minute. Her higher-pitched calls alternate with his lower-pitched ones. Her neck is bent slightly forward as though she is not at ease. His head is high and he scans all around. There is *no* gosling in sight. She swims over to her nest, climbs on, and then alternately looks down and gently picks at it with her bill, holds her head up and looks at me, and then picks and weakly pokes some more as though looking for her lost gosling(s). She becomes silent now, all the while her mate stays far away on the beaver lodge and keeps calling.

After twenty minutes she leaves her old nest and reenters the water, and then swims over to and around the nest site that she used (and abandoned) *last* year. She is searching, remembering, and not finding. After examining her two nest sites she swims across the pond to the south, meanders and calls among the sedges, and then crosses again while her mate swims over and joins her. She examines their recent nest a second time, and then both cross the pond to the beaver lodge. He hops on, and she stays alongside and dips her head down to throw water onto her back before they then both leave and keep calling, calling, calling endlessly.

An hour passes, and they eventually fly to the upper smaller pond, call some more, fly back, and as it gets dark finally settle on the beaver lodge. They leave early the next morning, and since they have now lost and not found their one gosling, I'm close to certain they will not be back until next spring.

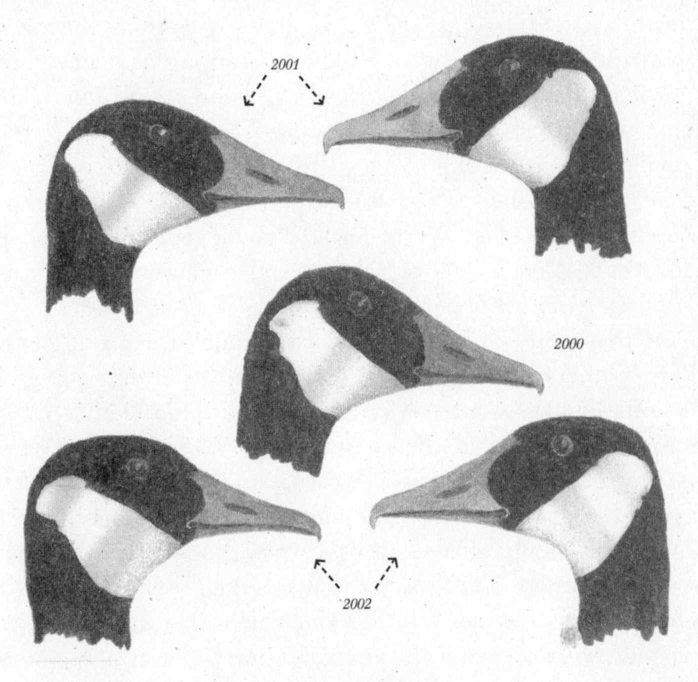

Jane in June over the years.

Jane and Harry were in all probability the geese who destroyed the Sedge pair's nest, and Jane, like Peep, had by now become a friend to me. In terms of the larger picture it probably makes little difference which particular goose smashed the Sedge pair's eggs to kill their about-to-emerge goslings. But to me it did matter, since I knew Jane as a "friend" and friends can disappoint.

The antidote to censure is understanding. How, then, can one understand a goose or a pair of them destroying the eggs or goslings of another pair, even as they adopt goslings and join up in communal superfamilies! What benefit can murder bring the perpetrators? I can think of one, and it is a consequence of the convergence of geese being long-lived, returning to breed year after year to the same spot, not tolerating close neighbors whose males might steal mates and whose females might dump eggs into others' nests, and having offspring with a tendency to return to the home area. That is, a neighbor this year may be a competitor next year. Since geese have a strong tendency to return to where they have nested successfully before, smashing the reproductive effort of a pair should induce it to nest elsewhere the following year. Smashing five eggs this year (after there is no more time for a second clutch) translates to five fewer potential nest spoilers in subsequent years. Such behavior could thus be adaptive, and if so, it could evolve, with specific emotions serving as the proximate motivators to induce "appropriate" action.

If that were all, however, then why didn't the geese go around killing others' goslings as well? Was it because other goslings could potentially be a benefit owing to the "selfish herd" effect? To have more "herd" around the geese should *encourage* others to nest nearby. At our beaver pond there were an infinite number of sedge hummocks, two beaver lodges, and two muskrat lodges to nest on. There was plenty of food. Another factor must underlie the intolerance of others, and I suspect that parentage trumps all where all else is available or not limiting. One clue is that intolerance wanes *on the nesting site*, after the young appear. On both sides parentage is potentially more in doubt the closer the next neighbor is to the breeding site. It may be advantageous for safety to be in a flock of either strangers or relatives. But matings occur only during and just before egg-laying. During that time the geese make sure no competition is near. Not only that, they act to make sure that none *will* be near in the future by expelling squatters who could come back to breed the next year. At the *feeding* site where the young are raised, however, breeding competition is not an issue, while safety is. Could this difference in shifting priorities be another reason for the separation of breeding and feeding sites?

16 ADOPTION, PARENTING, AND DESERTION

From our perspective, many animals make parenting look easy, but undoubtedly my conjecture is a consequence of my latest adventure with our current brood, Eliot and Lena. When I started my goose watching, the first was cutting his first tooth, the latter was not yet born. In the meantime, I've discovered all over again, as though recovering from an adaptive case of amnesia, that human youngsters require extraordinary and sometimes superhuman amounts of time, patience, effort, and resources—much more than two parents could reasonably provide. And, despite all that, I excuse myself for (to my wife) intolerably long hours chain-sawing and splitting wood, talking with students, preparing lectures, and so on; and watching the birds down in the bog, and maybe occupying a desk in the office, i.e., having fun. For me parenting is not a twenty-four-hour-a-day, full-time body contact tussle that continues day in, day out, night in, night out, without a break and with little variation. In desperation, respites with an electronic baby-sitter sometimes help to maintain sanity when meltdown is imminent, especially during winter months of near-total confinement.

We're envious of the geese. They do their parenting only in the spring and early summer. The young always stay together in a tight little flock, without any coercion. The family can stay together at all times. There is no need for any parent to be in two or more separate locations at the same time. And when the adults get tired and take a rest, the young immediately plop down to follow their example. There is little conflict between what one member of the family has to do or wants to do and what another might have to do, or wish to do. No messes ever need cleaning up in their primitive lifestyle. Modern human life has reputed advantages over that of our ancient anthropoid days, when the whole family spent their days on a series of continual romps through the woods and meadows of the great outdoors, but ease of parenting is decidedly not one of them.

Superficially it may seem that birds do it all with little fuss and no strain. The robin builds a nest in a day or two, lays several eggs, and sits on them for a couple of weeks. For two weeks both parents feed the eager young who unhesitatingly swallow all food put into their open mouths, and then it's all over. In geese and many other species, the parents don't even provide food. The young feed themselves from birth, and from day one they instantly respond to their parents' every signal to be quiet, to follow, or to hide. Many species of cuckoos in the Old World, and cowbirds in the New, even dispense with direct parenting entirely through the use of full-time baby-sitters. They insert their eggs into the nests of other birds who then do the job of egg-sitting as well.

Similar parasitism can also involve the same species, although then it is often caused inadvertently or accidentally, but it can also be a consistent parenting "strategy." Wood ducks and some other waterfowl, for example, often lay eggs into each other's nests, and such "egg-dumping" occurs commonly in waterfowl, especially when suitable nest sites (such as tree holes in the case of wood ducks) are few. With most perching birds whose young are born unable to feed themselves, the roles of the male and female tend to be similar: Both parents are required to provide for the brood (previous offspring sometimes help as well). However, when the young are precocious and can feed themselves, then the male no longer needs to be around to help gather food for his offspring. And generally with these birds the males aren't around. The number of offspring any female can have is then limited not by how much food she and her mate can bring, but by the number of eggs she can lay or incubate, while the male's reproductive output depends on the number of fertile females he mates with. The roles of the males and females, then, tend to differ greatly.

Sandpipers and some other shorebirds lay only four (huge) eggs per clutch because that is the number that just barely fit the female's brood patch (the patch of breast and abdomen often cleared of feathers and supplied with extra blood circulation to enhance heat transfer to the eggs): The four eggs have conveniently pointed ends to snugly fit together into a four-pack. But four is not necessarily the maximum annual output of young per female. In northern areas where food supplies are sufficient and summers short, some female sandpipers manage to rear more than one clutch at a time by "double clutching." To pull this off the female abandons her newly laid eggs. To ensure leaving offspring, her mate then has little option but to incubate and then guard the young. The female abandons both him and her clutch to find another male with whom to do the same thing all over again, secure that her first clutch is well taken care of. The northern golden-crowned kinglet, the world's smallest perching bird, which lays 8 to 10 eggs per clutch, double-clutches by a different mechanism. With this species the female abandons her just-hatched altricial (embryonic) *young*, but stays around and keeps her mate. The male then is forced to feed the young all alone while she gets busy and builds a second nest nearby that she fills with a second clutch of eggs. At that point her mate (to ensure his own reproductive output) must feed both her as she incubates *and* his first clutch of young. Males are used up to the limit of what they can and will do. And vice versa. In all of the diverse parenting strategies that these examples illustrate, the roles and responsibilities of the two sexes are not equivalent, and the best option is not necessarily the ideal one for any given mate. Each is a compromise.

With most precocial birds (such as ducks, quail, grouse, rails, many sandpipers) the males leave their mates and the nest soon after the eggs are laid. No long-term bond either with their mate or young is necessary to ensure a male's reproductive success. Simply put, once fertilized, the female no longer needs a mate because he can't contribute much to the welfare of the young. From her perspective, he might as well leave, and from his perspective, too, if leaving will enable him to father more young with another female. However, for male geese the option to leave is severely limited. Since the time slot for rearing young is very short (due to long developmental time required by the young and short northern summers), nesting necessarily takes place at the earliest possible time in the spring; hence it is highly synchronous within the population. There is thus little opportunity for serial mates, and this in turn has implications for parenting.

Parenting by geese is in many ways unique. With male perching birds like kinglets and sparrows, parenting success is largely contingent on the male being a superb provider, who can, however, offer little or no protection. With geese, the young feed themselves but can't fend for themselves. They are exposed to predators by running around at a tender age, and although the gander does not need to gather food, he *can* contribute by taking on a major role as protector, especially since he has little to gain reproductively by trying to acquire another female by the time his is incubating.

Geese are one of a few notable exceptions among bird species with precocial young in that they have a social system analogous to that of primates. The male parents among geese, swans, and cranes (who all have precocial young) are *especially* well-known for their long-term commitments to mate and offspring. How can one explain that anomaly? I think that with these birds it has to do with body size. Size confers the possibility of a credible defense, a role that most small precocial birds such as quail, grouse, rails, and others can't bring off. With small birds that cannot mount a credible physical defense, selective pressure for protecting the young has centered on other parenting options—hiding the young, inaccessibility of the nest, deceptive displays, and secrecy by the parents.

Aside from geese, only with some large hawks, eagles, owls, and swans does physical defense with the aid of the mate become a credible alternative. Who is not intimidated by a goshawk pair's clanging calls as the big, heavily taloned birds dive into one like World War II Stuka fighter planes, to rake their sharp talons across one's scalp? Who is not cowed by a big goose or cob who comes running at you, hissing threateningly and backing the threat up by slamming you over the head with its great wings?

A male goose makes himself indispensable to his mate by deflecting the predators' attention to himself and by his aggressive defense. He becomes the powerful escort that helps the young survive running the gauntlet of predators from the nest to and into their feeding pastures. He remains the family codefender while the babies are flightless, vulnerable, and visible as they grow up. His contributions allow the family to commit their attention to almost continual grazing on the open ground where their preferred food, such as grass, grows best.

I saw examples of the Canada geese gander's role in defense. In 2000, I once saw two domestic dogs (black Labrador retrievers) running into the water at Sunset Lane pond as though wanting to reach the geese. Each time they did so, one of the two ganders rushed over to them from mid-pond and successfully chased them back to shore. I saw another, more surreptitious, canine attack on June 25, 2001, a hot (90°F) and windless day. The two pairs with their goslings were swimming along the shore feeding on the seed-heads of ripening grass. The young hopped up into the air at intervals to reach these seed-heads. At other times they tipped their heads down into the water to feed underwater near shore. As the young were absorbed in feeding, I noticed that the gander with four now-gray, gangly "teenage" young, was nervous and alert and paying rapt attention to something on shore. Then I noticed a red fox in the grass at the pond edge. Occasionally it put its nose up into the air. Rather than retreating from this dangerous predator, the gander approached shore, to better observe or threaten the fox. Pop and Jane then swam over to join him. The four adult geese, followed by the fourteen young, drew closely together as they approached the fox, who then vanished from my view. I later saw it bounding off across the field. No predators succeeded in getting a single gosling at this pond during this, the previous, and the next summers.

When a threat approaches a goose family, the gander typically attacks and beats the predator with blows of his wings, or he stands tall in front of it and hisses threateningly, as happened when I approached nests with eggs where the ganders did not know me. The goose is tied to the goslings and cannot leave them to chase after predators. If one does get close, she stands her ground and spreads her wings to the sides to provide a protective shield for her goslings. She cannot leave them. But here, along the pond, such close encounters are rare probably because the open pasture allows the geese a long view, which gives them enough time to reach the safety of the water before such defensive measures are necessary.

With most birds who try to raise as many young as they safely can, the insertion of others' young into their own brood is reproductively equivalent to not living up to their own potential. Therefore, the adoption of unrelated young (and the other pair's abandonment of their own young), which I observed in my geese, was unexpected by me. What do those who practice such anomalous behaviors gain and what could they lose? There was no reasonable doubt about the occurrence of both the abandonments and the adoptions in my geese. The question is, how and why did they happen?

The adoption of strangers' young is generally defined by biologists as altruistic behavior, but when a bird like a warbler or a sparrow raises a young cowbird or a cuckoo, the phenomenon is not considered altruism but more appropriately "nest parasitism." That is because, although the adopters made a choice (even clever ravens accept in their nests young that they *recognize* as different), presumably they had been duped into it by exploitation of their inflexibly conservative innate behavior programs. They don't risk rejecting what *could* potentially be one of their own. However, perceptions and motives are irrelevant here. What matters are practical effects, as measured in terms of long-term evolutionary costs versus benefits by the yardstick of reproductive fitness. By that yardstick the correctness of the parasitism label is too obvious to need testing, since the cowbird or the cuckoo who inserts her young (via her eggs) into the clutch of another bird commonly breaks one of the other's eggs in her haste to dump the egg, or the young hatching from her egg, shoves the host's eggs or young out of the nest and starves the remaining young by monopolizing the food the stepparents bring. The hosts, in turn, evolve ever-sharper cognitive and behavioral traits that act to counteract the forced adoption. For example, they learn to recognize strange (i.e., different) eggs in their nests by their color patterns, and eject them from their clutch, thus eliminating the food hog before it arrives.

Adopting *precocial young* is, in contrast, at least not obviously costly on foster parents, because it occurs after the eggs hatch and the young can feed themselves. As a consequence, there is no or almost no competition among the young for food, and there is no readily apparent sacrifice by the adopting parents. There might even be an advantage to them.

A main limitation to the survival of precocial young is predation. However, by adopting the young of another family, a goose may be diluting the risk of having her own offspring eaten. For example, if half the clutch is from other parents, the host pair may be halving the risk of losing one of their own. But if adopting others' young is beneficial, why were the geese aggressive *before* the young appeared? Why didn't they nest next to others so that even more young "predator shields" could be easily acquired, and so that the dangerous journeys through the woods could be avoided? Why did the pairs of geese fight for the exclusive possession of the whole pond and the *whole* bog, even though there seemed to be ample nest sites and space and the shores were lined with green vegetation, the water alive with invertebrate prey and pondweed, and open fields nearby with clover and grass?

If one leaves out space and food as the contested resources, then the process of elimination suggests that the fights by the nesting or about-to-nest geese that I observed almost daily in the early spring at my bog had to do with ensuring *parentage*. Direct evidence for this hypothesis is the peaking of aggression just prior to and during the time of mating and egg-laying, the extra-pair copulations that occurred then, and a decline (and in some cases elimination) of territoriality that occurred after the young appeared. Intruding males could profitably mate with a resident gander's goose only during or near her egg-laying (sperm storage is less than a week), and ganders who through evolutionary time have been jealously defensive at the time that eggs are being laid were probably cuckolded less often than those who were then tolerant of other males in the vicinity of their mates.

There was hardly a single day all spring when the geese are laying eggs or during early incubation that one to three geese did not fly into my beaver pond either in the evening, at night, or in the morning or several times a day. These invaders were undoubtedly aware of the prominently visible female sitting on her nest on the muskrat or beaver lodges, because they usually landed near her. The gander always appeared in seconds and routed the visitors and he usually focused vehement attacks on the smaller, presumed female, of the invading geese. Why does the gander vehemently attack the *females*, rather than welcoming them? It was easy to rationalize that a gander "should" chase out males especially near egg-laying time to avoid possible cuckolding. But chasing females away when his goose is incubating on eggs seemed enigmatic. It made no sense—unless already fertilized females who have been unsuccessful in securing a nest site come to dump their eggs into others' nests. Egg-dumping is a well-documented reproductive strategy of numerous birds, especially in waterfowl (Sherman 2001).

A gander can no longer be cuckolded when his female is incubating, because "his" eggs have already been fertilized. And indeed *early* in the nesting cycle a resident gander focuses his aggression on other ganders. However, if there is competition for egg-space in the nest, then another egg added could mean an egg eliminated—one that *he* had fathered. Thus, he should try to chase off females that try to get near his nest, while his mate should stay on it to prevent opportunities for egg-dumping, to protect *her* maternal investment as well. The male's vigilance to potential egg-dumpers should depend on his female's tendency to lay as many eggs as she can successfully incubate.

Geese, like chickens, are indeterminate egg layers—the size of their clutch is not already determined when they start laying. Instead, they monitor their nest contents and keep adding eggs until their nest is full. Then they stop laying and start incubation. The Maine writer E. B. White mentioned (White 1977) how his domestic goose, Liz, produced forty-one eggs by late June, because he "continued to remove the early eggs from (her) nest, holding her to a clutch of fifteen." (He dated eggs with a pencil as soon as a new one was deposited.) Liz had, however, apparently greatly overestimated the number of eggs she could incubate, because only five of her clutch hatched. One can afford to be foolish if one is rich, and animals indulge in what they can afford to. Liz laid far more eggs than she needed or could incubate. E. B. White's pampered goose undoubtedly had unlimited food and could afford to be wasteful. But wild geese who work for a living are held to a strictly economical budget. They should try to produce all the young that they can, but they should never lay more eggs than they can incubate at any one time.

Adult wild Canada geese produce five to six eggs per clutch, and that is probably the right number. Clutches of up to nine eggs have been reported (Pearson 1917) although it is probably presumptuous to suppose that all of those eggs are their own. I don't know for sure how many more than six eggs a goose can sit on, but considering the size of the goose, and then the size of its huge eggs, five to six eggs must be close to the limit. (I took measurements of the clutch of six eggs of the Sedge pair; their eggs weighed 198 grams each, were 8.3 centimeters long and 6.2 centimeters maximum width, and the six covered a circle of 19.2 centimeters in diameter in the nest. In comparison, wild turkeys, who are larger than geese, have clutches of 10 to 16 eggs. A turkey nest I found next to the bog held 10 eggs, each weighing 73 grams, and that clutch covered a circle 17.5 centimeters in diameter.) That is, since goose eggs are large (the larger waterfowl eggs produce larger young that have a survival advantage over smaller young), few eggs already demand a relatively large area of the females' available brood patch space. It might be reproductively costly for a gander to permit even one egg dumped into his nest, and he should guard against it. Geese cannot be assured of their parentage of the eggs they incubate, unless they make sure other females don't dump eggs into their nests. Goose eggs all look alike, and eggs once laid are difficult if not impossible to discriminate from another's. Ultimately, geese contest *parentage*, much more than parenting.

The initial question that had puzzled me the most was why families immediately left the bog after their young hatched. In retrospect, this question became almost trite after I had watched the geese while they foraged and also acknowledged the possibility of predation during their forays onto the land, where they got most of their food.

Beaver bogs, because of beaver and muskrat houses and sedge hummocks, provide many mini-islands that are ideal nest sites. Surrounded by water at all sides, they afford protection from most predators. However, after the eggs hatch and mating and egg-dumping are no longer issues, the young have to leave these islets and the key issues are safety of the young and food. These two issues are linked.

Canada geese are tundra-adapted birds. If not tundra they require open tundralike areas such as suburban lawns, golf courses, and so on, where they can see into the distance. They need areas with short grass for grazing and for vigilance, especially since even the adults become flightless for a month during the summer wing molt. Safe areas with both food and long vistas are few and scattered in an environment such as Vermont that is dominated by forest.

Habitat was an important consideration in the geese's decision to stay or leave, because they stayed if the pond had short grass and open vistas surrounding it (Sunset Lane pond and Richmond-Hinesburg Road pond), and they left all the other ponds that are surrounded by forest and tall, dense vegetation. Sunset Lane pond was a particularly desirable site because it has no forest near it and is surrounded by grass that is kept short all summer by the cows that graze there. Predators there could not sneak up on a crowd of geese, even when most of them are preoccupied with their heads down in the grass.

An enigmatic question is why some pairs apparently deserted their young and let another pair adopt them. The answer to this question is linked to the above, because adoption is probably a consequence of the families' *simultaneous arrival* at one ideal foraging area that provides food and safety. When two (possibly related) families, each with *small* young that are not yet imprinted on their parents, come together, then these one- to three-day-old goslings initially have a strong tendency to follow almost *any* crowd. "Ownership" of these as yet unimprinted young would become blurred. The goslings could be shuffled between pairs—especially between relatives who know and visit each other. If the young follow the largest crowd, then one pair may end up with more than another. To the other pair it may then appear as though their young disappeared "into thin air." But even if they did recognize their young and wanted to retrieve them from the pack, the foster parents might rebuff any intrusion into what they perceive as "their" brood. That is, parents might not be physically able to separate their own out of the crowd. Meanwhile, mutual imprinting would quickly finalize the adoption by the adults of the young, and by the young "adopting" their new parents. For example, in 2002 at Sunset Lane pond, two sets of parents had 9 and 10 young, respectively. Two others had only 1 and 4, respectively. No family ever grew if its young were already about a week old, even though it might join up with another family to form a megafamily.

Why does the adopting pair accept the *other's* young? Practically speaking, adoption probably depends on the parents' inability to distinguish hatchlings, coupled with their unwillingness to reject any that could potentially be their own. Evolution could have produced a mechanism for more rigorous sorting out and early identification of foreign young. But I suspect that when there is little *cost* in having "extra" young, there will then be little selective pressure to evolve the ability to detect other young and repel them.

And why would the parents not evolve a mechanism to ensure that they *keep* their own young to themselves? First, they try to. But failing, they incur little cost: young are adequately cared for by the adoptive parents. Second, the biological parents may be freed to raise another clutch of young after their first brood is cared for. (However, in the north there is usually no time for that.) Alternatively and perhaps more realistically, geese without young are free to migrate north to the safe molting areas (in northern Quebec with this population). Perhaps the far northern areas are, or have been through evolution, safer areas for birds in the flightless condition; perhaps such summer migrants have been historically less vulnerable to aboriginal hunting pressure in the far northern and uninhabited tundra than in the teeming populated areas in the south. If being flightless in the south is more dangerous than being flightless in the north, one would expect a strong tendency toward summer migration to the safe molting grounds of the northern sanctuaries. However, the southern *breeders* can't afford to go there because they must stay with the goslings. Thus with their young well cared for and potentially even safer than before, both adoptees and adopters could gain reproductive benefits.

The geese's behavior looks as though it is based on intricate planning or strategic foresight. But the products of evolution that now exist are the *effects*, no matter how indirect, how long delayed, or how convoluted, no matter how seemingly illogical, or how unknown the outcomes. No logic is necessary to produce the behavior; we don't have to assume that geese think. Indeed, if they did think over such complex choices, they'd probably be even more confused than we are, despite our supposedly more complex brains.

Like the geese feeding on grass and the blackbirds on bugs, we ate proper food for millions of years without needing to know why we ate, with no knowledge of what and how much was healthy for us. We had no clue what happened to what we ingested or why we needed it. We were, and still are programmed with appetites and urges that induce appropriate behaviors, whether that concerns nutrition, reproduction, safety, or habitat preferences. Humans ate, slept, reproduced, made homes, fought, and made friends long before they had any notion of why they did it in terms of physiological mechanisms and consequences. Our passions were eventually increasingly nudged by reason, so they could less easily misfire as they commonly did with some animals in inappropriate contexts, such as Peep following a truck, redwings in California nesting next to parking lots and feeding on Chee-tos, a waterbug landing on a smooth dark pavement as though it were water, or some of us landing in jail. Like the birds', our behavior is nodded, prodded, and led by tastes, hates, thirsts, appetites, loves, fears, and jealousies. These predilections are parts of *basic* brain mechanisms. In science the simplest, most conservative approach is generally assumed to be the true one, until proven otherwise, and so far there is no proof that our basic physiology is built on a different scaffolding than a rat's or maybe a bird's! The emotions that drive our behavior are as physically bound to biochemical processes as are neuron, brain, and liver function. Such basic mechanisms are unlikely to be highly species-specific. That is, the behaviors that drive nutritional balance, reproductive success, and so on in geese are probably driven by emotions that are not alien to us and vice versa. What differs and differs *greatly* is our consciousness of them, and how the emotions are harnessed to different stimuli to achieve appropriate species-specific results.

17 A FEATHER DROPPED

My goose chase had become an unanticipated and fascinating adventure. I lived and recorded it willy-nilly as it occurred without expectation, and hopefully without bias, to try to reveal real patterns and isolate them from the imagined. There was, of course, the problem of when and how to end the book, since the geese live on.

The ending for the book came to me on September 2, 2002, more than a year before the end of the observations. It happened as unexpectedly as a thunderclap out of the blue. When I told Rachel, who had been in on the whole goose saga from day one, she said what I felt: "Nobody will believe this. They will just think you made it up. It's too perfect." Nature is seldom pat and never perfect. Least of all the geese. Diversity, ad hoc compromise, and the unexpected rule. What happened on September 1 was unexpected, but almost too artfully perfect. I was therefore tempted not to mention it.

You may recall that Peep had come back to us as if from the dead after an absence of two years. She and Pop, the gander she brought, had tried to nest but their nesting failed owing to other geese's attacks on the nest that killed three eggs, and at the end of incubation also their one gosling. They left the day after their gosling was killed. She came back without Pop the following spring and then bonded with Jack, a neighbor who had been a gander to Jane. Peep was expelled from the bog by Jane, who had bonded with Peep's former mate, Pop, in a mate switch. The next spring Peep came back once again. She was courted by another gander, and they were yet again violently expelled. And that, I was sure, was the end of the complex saga of our friend Peep, and it almost spelled the end of my interest in the social goings-on of the geese that I had come to know because of and through her.

The effects of the experience were indelible, of course. Post-Peep, I began to see geese in a different light. Every time I saw a goose, I thought of her. It could be her, or her gander, or Jane, or Harry, or one of the Sedge pair. Every time I saw a flock of honkers fly over, I felt one of my friends could be among them. It was as if I had been to a foreign land and had made a link to it through one of its inhabitants. I could now experience through them what previously I could not conceive.

As usual, by mid-August all the beaver ponds were silent, empty of geese. The goslings had again grown up and flown away. They were off with their parents to destinations that are to me unknown. The great Vs with hundreds of geese honking without pause as they fly on high toward the south were not expected for another month. I had finished watching geese, and I was no longer watching the bog at dawn. But during the night of September 1, I awoke with a start: one, or possibly two, geese were clamoring down in the beaver bog, and I was drawn to their calls by an unconscious and urgent pull. As dawn broke, I rushed down to be near them.

When I got to the shore, I saw not 1 or 2 but 12 geese on the far side of the pond. They were too far away for me to identify. However, I discerned that they consisted of two families, of 5 and 7 birds each. The two families stayed near each other yet slightly apart, like other families I'd watched team up for three summers at Sunset Lane pond.

"Hello—hello—" I hollered across the pond. The group of five ignored me. However, one bird in the family of seven raised its head high and started to paddle toward me. The other six of her group followed. As the one goose came closer, I heard baby-goose peeping from her followers. So she was the parent of a family raised this year. This family with five young had a thick-necked gander with an unusually narrow white face bib bringing up the rear. I had not seen him before.

The lead goose seemed to recognize me and was eager to approach. It was not Jane, though, because the gander was not Harry, and Harry and Jane had apparently left their six young this spring with the Sunset Lane pair. They would not have retrieved them. When the mother of the five now adult-appearing young got still closer, I saw that her face bib matched Peep's. Could Peep have raised a family and come for a visit?

"Peep—Peep!" I repeated. The goose raised her head again, and again she and her family paddled closer. However, the sedges and cattails were tall now at summer's end, and they separated us. The family of geese swam up to the edge of the green wall, loitered there, and started feeding. Only the mother goose kept her head high, still looking at me. I threw out a handful of grain. Again she edged a bit closer, advancing into a small cove of open water, but I still could not identify her.

All summer I had wondered what happened to Peep after I had seen her meet and apparently bond with a narrow-bibbed gander at this, her home pond. To have her now return with him and five young was wishful thinking. It was too much to expect, and I did not allow myself to believe it was her.

The goose kept looking but stayed away. The grain did not tempt her and I was disappointed that I could not get close enough to verify her identity. And then both the gander and her young started pulling back in the opposite direction, and she followed them. The dense pondweed separated us. If I waded out into the pond, all the geese would fly off. . . .

I stopped to watch in disappointment from a distance as the seven geese returned back out into the middle of the pond to rejoin the other pair with their three young. I recalled then how Peep had returned the previous fall, all by herself, making long sad-sounding and agitated calls as though she was discontented and searching for something. Now I was searching and discontent. I wanted to think her life was fulfilled. But after I walked through the woods back to the house, I realized that nice thoughts come easy and they can seductively subvert reality.

I sat at my desk and started to write up the frustratingly inconclusive observations. I had barely begun when I again heard the group of geese starting to honk in the bog below the house. Knowing they were getting ready to fly off, I dropped my ballpoint and rushed outside to try to see them head down the valley where the geese usually fly when leaving the bog.

To my surprise, I saw the group of twelve geese flying directly up toward and just beyond the house after having come up the hill over the woods. Only Peep, and her previous mate Pop, had ever taken that route before. I yelled, "*Peep!*" The lead goose made a U-turn, the others followed, and then in a wild loud clamor they came in my direction. I jumped off the porch and ran down onto the lawn. The group made another turn, and then she set her wings

Peep coming in for a landing with the group following behind her.

and lowered her feet and started gliding down through the air directly toward me. She came right by my head and was about to land just beyond me, but the rest of the flock behind her then banked up and out over the trees. Then she backpedaled as well and I heard the heavy pounding of her wings as she barely missed hitting the wall of trees on the other side. The exertion of her overweighted wing-beats while turning sharply and trying to regain altitude in coming out of the clearing dislodged a wing feather. As Peep rejoined the flock, the feather drifted down and settled practically at my feet.

I'm a confirmed empirical rationalist. But as I bent down to pick up that feather, I could feel myself shake involuntarily. Peep had made a life for herself after all, and she had flown out of her way to me for no practical purpose. I, too, had reacted from deep emotion. At the start, her coming into my life had been a random act of fate. Now, her dropping the feather was another. However, it felt more like Peep was leaving a token of our friendship.

The feather that dropped from Peep.

About the Appendices

During my research I tried to resist the scientist's impulse to squeeze Peep, Pop, and the other individual geese into a larger theoretical framework. And while I do reference scientific findings throughout *The Geese of Beaver Bog*, in the field observations of these special geese I've tried not to interrupt the narrative. Instead, I've included a series of appendices.

Appendix 1, "Chronologies at Two Ponds," provides a timeline of major events. Appendix 2, "Lorenz's Geese," provides an overview of Konrad Lorenz's study of greylag geese, focusing on elements that are relevant to the Peep story.

In Appendix 3, "Canada Goose Populations," I face a quandary: how to incorporate hunting into my discussion (geese have been hunted for thousands of years, so one cannot discuss their populations without it) without making my portrayal of Peep seem like antihunting propaganda. Many northern people's livelihood depends on hunting, and I personally find it more humane, and more honest, to hunt wild birds from a population that is at carrying capacity of the environment than ordering chicken from the drive-through window at McDonald's. I eat some meat, and I know where it comes from. I can't climb onto the moral high ground and say I don't approve of someone eating wild geese. If I ate only corn (where "I" plow the prairie and destroy habitat for geese, ducks, etc.), would I be any more innocent?

I acknowledge that individual geese have a right to life, as do hawks, and owls, and mice. Rights conflict. There are rights beyond those of the individual. In the world of living organisms we must refer to a higher authority than the individual. Ourselves included. In terms of practical reality, that's the population, the ecosystem, and ultimately, for the geese, the bog.

Appendix 4, "Birds Seen at and around the Beaver Bog," provides information on other inhabitants of the bog, who so often provide the soundtrack to Peep's story.

And finally, for the reader who may want to read more: There is a vast potential literature and in my list of references I make no pretense to try to be complete. The geese are not just one topic. Their lives and their interactions with their environment concern *all* of biology! To select is, in effect, to elect to leave out, but *The Geese of Beaver Bog* is not a review paper and so I have provided a list of references, grouped by subject, hopefully helpful—but certainly not exhaustive.

Appendix 1: Chronologies at Two Ponds

THE GEESE ON MY BEAVER BOG

1997. The first (?) pair of geese are found nesting on the bog. Nest on sedge hummock. Four eggs. Three young left.

1998. The (same?) pair nest on an old beaver lodge. Five eggs. Five young left. Peep is born.

1999. The same pair nest on same beaver lodge. Five eggs. Five young left.

2000. Pop and Peep attempt nesting on sedge hummock in upper pond. Four eggs. A *nonnesting* pair stay, and destroy Peep's nest. The resident pair (now called Jack and Jane) abandon the beaver lodge (active this year) and nest on the muskrat lodge. Six eggs. Four young left.

2001. Pop and Jane nest on a muskrat lodge. Six eggs. Five young left. Peep returns alone and is driven out, leaving with Jack.

2002. A new young pair, the Sedge pair, attempt to nest (inspecting numerous sedge hummocks and a beaver lodge) and settle on a mound of debris. They lay three eggs, briefly incubate them, and then abandon the nest. Harry and Jane nest on the abandoned beaver lodge. Six eggs. Six young left.

2003. The Sedge pair come early and nest on the muskrat lodge. Six eggs. Five of their ready-to-hatch eggs were destroyed by Jane and Harry, who the Sedge pair had harassed and dominated. The Sedge pair lost their one gosling after leaving the pond.

THE GEESE AT SUNSET LANE POND

1997–1999. ? Residents unknown.

2000. Resident pair raise five young, joined by another with three young. (Eight young raised to adulthood.)

2001. Pop and Jane come with five young and adopt five more from another pair originating elsewhere. Another pair come with four young and a total of fourteen young are raised to adulthood. A young pair lay four eggs, briefly incubate, and then desert the nest.

2002. Resident pair with five young increase their family to nine young by adoption. Harry and Jane (?) leave six young with another pair that came previously with four young. Simultaneous arrival of two other pairs, one with four and the other with one young. (A total of twenty-four young of six pairs, none of whom nested at the pond, are raised to adulthood.)

2003. A new and aggressively defending non-nesting pair stay until late May. After they leave, two pairs with four and five young each come, to make a superfamily of thirteen geese. They are then joined by a third pair with one young. The sixteen geese stay together as one "family" all summer until the young grow to adulthood. They then leave together in late August.

Appendix 2: Lorenz's Geese

After my adventure with Peep and her associates, I was curious to find out how my observations fit in with what was known about goose behavior. Previous to that I was reluctant to read about what I preferred to witness firsthand. Konrad Lorenz and flocks of students had spent a half century studying the greylag goose *(Anser anser)*. Collectively their work spanned over a century. It is far beyond my scope to review it all, and this book is not meant to be an authoritative treatise about geese. Excellent accounts of geese have been written already, and one of them is Konrad Lorenz's 1988 *Hier bin ich—wo bist du? Ethologie der Graugans* (Here am I—where are you? Ethology of the greylag goose). I had received as a gift a copy of this book from the Grünau Research Station in Austria where Lorenz did his famous studies on the greylag goose, where the descendents of the flock of geese he introduced still exist, and where students continue to work with them. If anyone ever had a grasp of goose behavior then it should be Lorenz, and I therefore read his book, the sum of a lifelong research into the greylag goose. I took notes and was tempted to insert references to this work throughout my narrative. However, I felt that would disrupt the flow of my observations, so I decided to leave them unencumbered, and instead here give a selective overview of greylag goose biology that seems relevant and/or interesting to the Peep story and Canada geese.

As a child Konrad Lorenz was read a book by Selma Lagerlöfs about wild geese that fired his imagination. Shortly thereafter, at age six, his mother bought him a duckling (whom he named Pipsa, pronounced "peepsa" in English) against the objections of his father who thought that to do so was a form of animal torture. (I had no knowledge of this when I named Peep, and when Eliot inherited her over the objections of *his* father.) However, the duck thrived and lived to the age of fifteen years. Lorenz's early experience with the duckling "imprinted" him on waterfowl, and later as an adult he adopted a greylag gosling, Martina, who was hatched under a domestic goose. She became his companion, and over a year even slept with him in his bedroom. As Lorenz wrote in his 1952 classic *King Solomon's Ring*, "It is only by living with animals that one can attain a real understanding of their ways." His intimate living with Martina made possible "eine Reihe von Zufallsbeobachtungen" (a series of chance observations).

His "chance" observations led to questions, new observations, and ultimately to understanding. I will describe the biology of the greylag goose as revealed to him and his associates, concentrating on what may also be relevant to Canada geese, starting from the egg stage through hatching, to adulthood and pairing and nesting.

The greylag incubation period is, as in Canada geese, 28 days. For birds, this is a relatively long incubation. For example, grouse and chickens hatch after 21 to 25 days, and most small songbirds after about 10 to 12. At least two variables contribute to length of incubation: egg size and degree of development of the hatchlings. Larger young take longer to develop, and while perching birds are still virtual embryos when they crack out of the eggshell, geese only emerge after they are covered in feathers and equipped with a large portion of their instinctual behavioral repertoire.

Much relevant behavior is already being executed before the goslings emerge from the egg. While still within the egg, they communicate with the parents using three kinds of vocalizations (Lorenz 1988, p. 123). First, there is the "Weinlaut," analogous to a human baby's cry—a high-pitched long-drawn-out monosyllable. Lorenz also calls it the "Pfeifen des Verlassenseins" (piping of being left or abandoned). This call is given in response to any acute discomfort. The goslings give it when the egg cools, and it induces the mother to brood. Hatched goslings are unable to regulate their body temperature, and give the same cry to induce their mother to brood them. The goslings also cry when they encounter difficulties escaping the eggshell, or when being hindered by an obstacle while trying to follow the mother. As in crying in other animals, the call induces the mother's solicitations.

The second call, used both inside the egg and outside, signals well-being and contact. This multisyllable call starts when the cooled egg becomes warmed (Lorenz 1988, p. 123). If a human talks to the egg, "it" also responds with these comfort calls. The third call is a "trilling" that is given when the goslings go to sleep, and when they are brooded by the mother. The goslings in the eggs are also in communication with each other, egg to egg, to synchronize their hatching.

The act of hatching itself involves a specific behavioral repertoire in which the gosling presses outward against the eggshell with the egg tooth (a real tooth with enamel) at the tip of its rostrum—the top of its bill. It never pecks. The bird rotates about within the egg to loosen the eggshell in a circle at the thick end of the egg.

The just-emerged gosling looks matted and wet. But it is less wet than it appears since the strands of down feathers are pressed together by being enclosed in thin, horny sheaths. These sheaths are worn off through rubbing as the young become active under the belly of the mother. This rubbing results in "electrical loading" of the feathers (Lorenz 1988, p. 158), which causes the thin down plumes to repel each other and become uniformly spaced so that the gosling becomes fluffy, as is required for waterproofing. Preening by adults serves the same function, and it is analogous to our hair combing. The static electricity generated by mechanical activity thus explains why the goslings don't leave the nest until one to two days after hatching, and perhaps also why adult geese spend so much time preening.

Lorenz (1988, p. 55) claims that the mother greylag gives the signal to leave the nest, although in the case of Jane and Harry, it appeared to be the gander who took the lead and then the mother led the goslings. At first the goslings gather in a thick clump, then follow each other in a chain. With greylags there are often big fights as young from one nest join the brood of a nearby nest hatching at the same time. The pair that loses the fight, and thus their young, follow the double-clutch pair for about a day. Lorenz also notes that the parents often lead the young a long way from their original nesting site.

If geese have been made famous for one thing, it is surely the concept of "imprinting." Imprinting is defined as a learning process that is quick, irreversible, and occurs in a very circumscribed window of time during development. With geese, imprinting occurs during the first couple of days after hatching. At first the goslings follow any moving object that makes a rhythmic series of sounds in a certain frequency range. Since, at least in Canada geese, goslings just off the nest follow only the mother (whom I heard making rapid, low-frequency grunting sounds), I presume that the gander is silent or they would follow him and her indiscriminately. The goslings quickly become imprinted onto the species, but when leaving the nest they are not yet fixated on particular individuals of their kind. Indeed, imprinting refers to *species, not* individuals within the species. Within about twenty hours of hatching the goslings do not yet recognize their parents as individuals. Lorenz (1981, p. 275) points out that if after the age of two days a gosling tries to join up with a strange family with goslings near the same age, their parents will bite it. Thus, apparently parents also soon learn to recognize their own offspring, even as the young learn to recognize them.

How does individual recognition occur? Lorenz (1981, p. 275) believes that the goslings at first identify their parents by voice. Young greylags usually stay with their parents for at least a year, and learn to recognize them in the same way that humans recognize individuals: by their faces. Hand-reared goslings treat humans as individuals as readily as they do their own species, by transferring their innate powers of discerning facial features of their own kind to humans.

Geese do not display their behaviors in isolation. They are not real geese at all unless they are with others; they are social animals. Their connectedness with their fellows extends to all aspects of their lives. Even flight. Lorenz (1988, p. 146) thinks that the famous V formation in flight functions to give the geese a "freien Blick" (unobstructed view) to the front. I presume it does provide such a view, but he presents this hypothesis as an alternative to a possible aerodynamic advantage of the following goose. "Ganz sicher zieht die nachfolgende Gans keine aerodynamische Vorteile an der Flugarbeit der vor ihr fliegenden." (It is certain that the following goose does not receive any aerodynamic benefits from the one flying in front of it.) However, recent experiments prove that the following bird in V formation does indeed gain an energetic advantage (17 percent reduction in energy expenditure) from the flight work performed by the bird ahead of it. The leading bird creates a turbulence wave that gives lift to the one behind.

The maximum observed longevity of free-living greylags in the field in Lorenz's studies was about twenty years, and 50 percent of males and 75 percent of females get widowed. Although geese who have lost a partner (males sometimes pair with males, but females never with other females) show all the symptoms of grief seen in humans, they may find a new partner in days. Hand-reared geese who lose a partner often return to the proximity where they were raised. Adult young who lose parents also show grief. They call and look constantly, "optimistically" approaching all geese who have their heads under their shoulders (and who are at first not recognized).

Greylags may fall in love "almost instantly" at almost any time and regardless of their reproductive cycle. Pair-building is initiated by the male, and a male who falls in love is "plötzlich verwandelt" (suddenly metamorphosed) into a different being. He shows increased courage, strength, and aggressiveness and initiates "Imponieverbalten" (behavior to impress) to his desired female, and tries to keep rivals away from her. If he is already paired, his wife will wait until he has initiated his display ceremony in front of his new love, and then at the critical moment she will insert herself between them so that he ends up performing it to her instead

(Lorenz 1988, p. 277). I suspect this is the "confusion" I saw when Peep and Jane were in conflict over the same gander.

The end result of the courting that seals a marriage is mutual "Triumphgeschrei" (triumph yells) between partners. The triumph display is the most important factor keeping pairs together. It was already well described by Oskar Heinroth, Lorenz's "doctor father" in 1910, as quoted by Lorenz (1988, p. 246) and here translated by me:

"A goose pair . . . comes in the vicinity of a stranger of its own kind. The gander then runs or swims with stretched-out neck angrily at the stranger, and this bird then generally retreats. Immediately the attacker then makes haste to return to his mate—and then both begin the triumph yells and ceremony."

Lorenz (1988, p. 244) gives further description of this important display: The gander opens his wings to the side, chest forward, and head down in greeting to the female. She responds by coming to him and also lowering her head in greeting, during which there is constant vocal exchange. During this time, they "both look as though they would fall all over each other, and they yell their loud calls directly in the ears, while keeping their heads close to the ground the whole time." The male's show of pairing readiness is simultaneously a demonstration of his maleness, although in long-married pairs, the macho show declines.

The repertoire of possible behavior that he devotes much of his book to is what is called the "Ethogram"—a list of all that a goose may or could do. Males and females have not one movement pattern that is gender-specific. What varies between them is the *frequency* of the behaviors, and how they are activated differently by hormones and circumstances. Many of these instinctive genetically coded behaviors or "fixed action patterns" (FAPs) are released by specific stimuli, called "key stimuli" because they are analogous to a key that can be used only on a specific lock to open a specific door.

One example of a FAP is the egg-rolling behavior studied primarily by the Dutch ethologist Nikolaas Tinbergen. In this behavior a goose (or a seagull) retrieves an egg that has rolled to the edge of the nest by reaching forward, bending the head down and applying pressure to the egg and rolling it back to the nest by balancing it along the underside of the bill. If one snatches the egg away after the behavior starts (has been "released"), the goose continues it as though rolling a phantom egg (Tinbergen and Lorenz 1938). This is not to say, however, that learning can't superimpose and modify a FAP.

With geese, only the goose (never the gander) incubates and broods the young. Normally, that is. Lorenz describes how on one occasion a female greylag was killed by a fox. But the young and the gander remained. The young then tried to get under him for shelter. At first he didn't lift his wings to let them in, but gradually learned to do so. Most interestingly, when he married a new goose the next year he shared the brooding responsibility with her (Lorenz 1988, p. 211).

Lorenz found that the greylag ganders were "never" near the nest during incubation, and they only came to the nest near hatching time. He acknowledged that he did not know if the gander then comes because he hears the young or because he is informed by the mother. My Canada geese ganders behaved similarly, but I don't have a clue as to how the gander came to be at the nest around hatching time. Could they measure the time of incubation?

With the greylag, love can easily turn to hate. Specific individuals may be singled out. Some ganders are caught in a constant conflict situation as if bound to it. Some fights are nearly to the death, as between his ganders named Markus and Blasius. Hate can also be transferred to specific humans. One pair of brothers, Keit and Rufus, took a particular hate to Lorenz's assistant, Paul Winkler (Lorenz 1988, p. 284), possibly because "in their sensitive youth they received some impressions of him being an enemy." These two, once adult, attacked not only his person, but his car, sometimes even when he was not in it.

Lorenz's long involvement with geese, especially in the area of pair formation, gender division of labor in parenting, mutual friendships, jealousies, hatreds, and vocalizations, gives much that is intelligible to humans. He is impressed by the "unbelievable analogies" between humans and social birds such as geese, where father, mother, and children have long-lasting, sometimes lifelong, bonds.

There is no doubt that scientists who study geese and other waterfowl are indebted to Lorenz for his dedicated research and detailed observations. Unfortunately, there is a dark side to the Lorenz story: his entanglement with National Socialist ideology. It has been discussed elsewhere in greater detail (see Kalikow 1980, Deichmann 1996, Klopfer 1999), but I do want to acknowledge it here and to do so I first need to provide some background on behavioral science.

Lorenz, as one of the "fathers" of the behavioral science of ethology, emphasized that the behavior of many animals is genetically transmitted through lineages, making it a marker of phylogenetic origin similar to such other species-specific identifying characteristics as fur and feather coloration. An animal's behavior is not only a function of learning. It is also explicable in terms of the environment to which the animal is adapted, as well as its evolutionary descent. (The "environment" includes not only the physical aspects of terrain and temperature, etc., but also the social conditions within which the animal lives.) Given the specificity of adaptation, Lorenz argued that hybridization would break up genetic coherence and result in harm of the species; it would allow respectively adapted genomes to mix, resulting in inappropriate behavior. Following the trajectory of this logic, Lorenz went on to say that under the conditions of domestication (and civilization in humans), animals were freed of the stringent selective pressures that pruned variants out of the population in the wild. The species thus becomes variable in appearance and behavior. Indeed, Lorenz hypothesized that under the umbrella of domestication/civilization, the "deviants" who would normally be selected out would gain a reproductive advantage over the wild genotype.

These ideas were not particularly novel. The problem was that Lorenz injected values into his science. He adhered to the then-current (and now long-discredited) notions that animals behave or should behave for the "good" of the species and that individuals who differ from the population norm are "deviants." Lorenz believed that hybridization or a relaxation of selective pressure would "show symptoms of decay" that needed to be vigorously counteracted. Such appeals were particularly harmful because Lorenz combined them with the error of not distinguishing species from "races." He rejected empirical evidence in favor of theory and extrapolation, and imposed his ideas about waterfowl on supposed human "races," which, if "hybridized," would become degraded.

It gets grimmer. Konrad Lorenz willingly joined the Nazi party (he was rejected on his first application), and served in the Rassenpolitisches Amt, the race division of the SS, as a psychologist in Posen. He performed tests on German-Polish "hybrids" (such as myself) to evaluate their "worthiness," determining who would be sent to the concentration camps. In a paper written in 1940, two years before his assignment, Lorenz articulated the theory behind the "unworthiness" idea. In that paper he explains that "precisely through their inferiority" (for having lost the "good for the species" traits) can such hybrids and other unworthies "invade the healthy body of the nation, and eventually destroy it." He made still another analogy with further value-laden extrapolation that left little to the imagination: "Just as with cancer-suffering [for which] mankind cannot give any other advice than to recognize the evil as soon as possible and then eradicate it, so racial hygiene defense against elements afflicted with defects is likewise restricted to employ the same primitive measures."

Appendix 3: Canada Goose Populations

Eleven races or subspecies or populations of Canada geese have so far been recognized (Bellrose 1976). Birds of these groups vary in size from as small as the 2.8-pound (female) and 3.4-pound (male) cackling Canada goose, *Branta canadensis minima*, to the 11.1-pound (female) and 12.5-pound (male) giant Canada goose, *Branta canadensis maxima*. Bill lengths vary from about 28 millimeters in the first subspecies to near 60 millimeters in the latter. Neck size and darkness or lightness of pigmentation differ as well. Many birds do not easily fit into designated races either because there are still more races or there is interbreeding among the populations. All of the eleven officially designated races have traditionally been long-distance migrants, and most breed isolated from each other primarily on the tundra.

The geese in the northeastern United States and Canada were originally the North Atlantic *B. c. canadensis* and the mid-Atlantic *B. c. interior*. Together they constitute the Atlantic Flyway geese. The first race breeds through Labrador and Newfoundland and winters along the Atlantic coast all the way south to North Carolina. The second one breeds primarily on the tundra of the Ungava Peninsula in Quebec (an area that is also used as a molting ground by the two other races of the Tennessee and

Mississippi Valleys). These two races constitute the migrants that we see here in the East every spring as they fly over very high on their way to the north to breed, and then when they come back in the fall to overwinter in the southern United States.

In pre-Colonial times there may also have been resident *B. c. canadensis* breeding in the eastern United States but if so these were extirpated. The ornithologist T. Gilbert Pearson, writing of the Canada goose, said: "The great breeding grounds of this goose are in the British provinces. . . . [In winter] these birds assemble in enormous numbers on favorite feeding grounds in Chesapeake Bay and in the sounds of North Carolina. In Currituck Sound I have seen one flight [formation] that was two hours in passing a given point. They came in one long wavy rank after another, [with] twenty and thirty of these extended lines of geese being in sight at a time." (Pearson 1917) There is now a thriving and still rapidly growing population of resident Canada geese within the Atlantic Flyway and these geese are not of pure North Atlantic Flyway stock. They exist largely because of recent human agency—the creation of open tundralike habitat that has replaced forests. They are a separate population from the geese that continue to migrate along their historic flyways to Labrador, Newfoundland, and the Ungava Peninsula in Quebec.

The present-day resident (breeding) population in Vermont and all over the eastern United States is comprised of a mixture of *B. c. maxima*, *B. c. canadensis*, *B. c. interior*, *B. c. moffita* (a western subspecies), and possibly others. These birds were derived from captive flocks that were at first kept by private individuals, often for use as live decoys to lure in the migrating *B. c. canadensis*. Captive flocks were then liberated after 1935 when the use of live decoys for hunting was prohibited. State wildlife agencies also started stocking programs in many states of the Atlantic Flyway, and resident goose populations boomed, both in numbers and in geographical distribution. Now "Canada" geese also exist as feral populations after introductions to Great Britain and New Zealand.

Resident Canada goose populations increased dramatically since 1935 even after stocking programs were terminated in 1990. For example, in New Jersey, goose populations have increased from about 2,000 to 240,000 since the early 1960s. Until 2002 wildlife officials in New Jersey, New York, and Pennsylvania have captured geese and sent them to Mississippi and Arkansas. Now, no state accepts them, and the U.S. Fish and Wildlife Service has the ambitious goal of slashing the resident U.S. goose population by 50 percent. The northeastern United States populations still continued their phenomenal population growth of about 15 percent per year. At one time the giant Canada goose was thought to be extinct, but it and all the races are thriving now and the increase in the Canada goose populations is a spectacular achievement of wildlife management that is equivalent to the comeback of white-tailed deer and the resurgence of wild turkeys. Resident Canada geese are thriving all through New England, New York, New Jersey, Pennsylvania, Maryland, Virginia, and south to Georgia. Their breeding range is extending both south and north, and it is projected to someday merge with that of the migrant geese from Quebec (Bellrose 1976).

There were no resident Canada geese in Vermont in 1956, when 44 wild-trapped geese were released on the Dead Creek Wildlife Management Area in Addison County. This site is about thirty miles southwest of the beaver bog that was the focus of my observations. As already indicated, the geese using one of the nearby ponds, Sunset Lane pond, as a nursery increased from 12 (4 adults) in 2000 to 18 (4 adults) in 2001 to 32 (8 adults) in 2002, but in 2003 they were again down to 16 (6 adults).

As in all resident goose populations, the precise racial background of the Vermont population is unknown. The 44 original birds at Dead Creek came from Bombay Hook National Wildlife Refuge in Delaware, and these birds are thought to have been progeny of captive birds released by private landowners. Other resident birds in Vermont are suspected to have originated in Massachusetts or New York and are also descendents of birds once kept on private estates, and they probably include *B. c. canadensis*, *B. c. interior*, and *B. c. maxima*.

My personal bias in trying to understand the behavior of Canada geese is to focus more on cultural differences than racial stereotyping. The resident-versus-migrant dichotomy may represent distinct genetic tendencies created by artificial selection in breeding pens. But I doubt it. Canada geese migrate by following their parents. That is, if one were to switch eggs between the migrant and nonmigrant population, the goslings would follow the flocks of their adoptive parents in the same way that Peep followed me and my truck.

Canada geese may seem overabundant to some people. However, it is precisely because of their abundance that they have had a key role in the ongoing success story of whooping crane conservation. Canada geese gave us the first clear protocol of what needed to be done, and in a very surprising and counterintuitive way. It started with the Canadian artist, pilot, and naturalist William Lishman who, following the example of Konrad Lorenz, imprinted geese, but on an ultralight aircraft. He trained two separate flocks of geese to follow him in his ultralight from Ontario, Canada, to Virginia and South Carolina. To everyone's great surprise and joy, the geese returned home to Ontario on their own in the spring. Lishman's amazing trip with the geese inspired the Academy Award–nominated film *Fly Away Home*, a fictionalized account of Lishman's life that ushered in the amazing conservation story of the whoopers. If it had not been for the geese, who were so common as almost to be a "nuisance," no government agency would have had the confidence to allow anyone to tamper with the majestic and critically endangered cranes. As a consequence, nobody would have learned about them so intimately, and the cranes' chances of success would have been drastically reduced.

George Archibald of the International Crane Organization noted Lishman's results with geese. He approached him to try to experiment with sandhill cranes as stand-ins for the whoopers, to see if the results might eventually be applied to the long-standing dream of establishing a second migratory flock of whoopers. The only existing one was that which had been reduced to the very brink of extinction, at 16 individuals in 1941. This tiny flock has been carefully tended and it has now grown to 184 individuals. It annually migrates the twenty-four hundred miles from its overwintering site at the Aranas National Wildlife Refuge in Texas to its breeding area in the remote Canadian north of Wood Buffalo National Park. That single flock could easily be snuffed out if caught in a hurricane or some other disturbance.

In 1997, Lishman and pilot Joe Duff imprinted sandhill cranes, as stand-ins for whoopers, to follow their ultralight aircraft, essentially duplicating what they had done with the geese. This second success set the stage for the agencies to risk granting the most relevant but also the riskiest experiment. In 2001 the first (trial) flock of eight captive-bred whoopers left Needah National Wildlife Refuge in central Wisconsin, guided by an ultralight aircraft, and headed on a journey twelve hundred miles distant, to the Chassahawitzka Wildlife Refuge in Florida. To immense relief and exultation of everyone involved, the birds all arrived safely. Two later became victims to bobcats in the park, and one was electrocuted on an electric power line, but five of the birds survived the winter. Those five then successfully returned *unguided* over the twelve-hundred-mile route back to Wisconsin. In the spring of 2002, the Wisconsin-Florida flock had grown to twenty-one birds. A new migration route, and a new breeding flock of whooping cranes, has been established. Thanks to a wild idea, to Canada geese, and to innumerable dedicated people with a dream, our descendents may one day be thrilled again by the spectacle of the whooping crane.

Like the cranes, geese are not born knowing where to migrate. They normally become imprinted on a migration route by following their parents and other family members and their associates. Consequently, if one raised goslings of the migrants in pens, they would have no migratory destination, although they could still disperse when necessary as resident birds do now. They would expectedly act like residents because the cultural heritage of their migration routes and destinations would be obliterated.

Some flocks of the present resident geese, although reluctant to leave their breeding areas, do move up to several hundred kilometers from their breeding areas if necessary to find open water and feeding areas in the winter. Additionally, as already mentioned, subadults and other nonbreeders migrate to northern Quebec in

May and June (into the breeding areas of the migrants) to molt their flight feathers.

Genetic distinctiveness is apparent among different populations, but species designation is not so clear. Waterfowl are notorious for being able to hybridize, even across distinct species lines. Indeed, there is concern that the black duck, *Anas rubripes*, might become locally extinct in some areas as a result of hybridization with mallards, *Anas platyrhynchos*. Mallards can also hybridize with pintails, *Anas acuta*.

Canada geese form many distinct populations because, despite their often long migrations, they are highly site-specific for breeding. They come back home to the turf where they were born, so genetic differences can arise by chance in small founder populations and then accumulate if not swamped out by interbreeding into a larger population. As a consequence of the strong home site fidelity, local adaptation, and little genetic mixing is expected even without genetic barriers. I suspect, therefore, that subspecies classifications, although they apply ecologically, would provide no breeding barrier if one were to switch eggs around. In effect, that experiment has been done: The resident populations are derived from several populations that now interbreed indiscriminately.

Given the advantage of not migrating due to less hunting or lower predator presence, the increasing availability of open water and year-round food in urban environments, and enough time, resident populations would undoubtedly eventually have become established on their own. Human introductions that have spread geese around have sped up the process. Humans created an open niche in the fields, farms, lawns, golf courses, ponds, and parks, and the birds are now filling it.

Subsidizing geese populations through human agency can have far-reaching consequences. Take the example of the snow goose, *Chen caerulescens*. In the past thirty years, in response to the use of nitrogen fertilizers in agriculture, these geese have moved inland from the salt marshes along the coast of Texas and Louisiana where they traditionally overwintered to feed on crops such as soybeans, corn, and wheat. As a result, their populations have increased "by at least a factor of five" (Bertness et al. 2004). That increase has, in turn, seriously disrupted a habitat 5,000 km farther north in the Hudson Bay lowland's coastal marshes, where snow geese breed in large colonies. There, one goose removes all vegetation above and below ground at about one square meter per hour, and with an estimated population of 3 to 6 million birds, snow geese are increasingly turning salt marshes into mud flats, and they are also causing a sharp decrease in the productivity of other arctic habitats utilized by seals, fish, and polar bears.

In parts of North America much farther south, resident Canada geese are reputedly already exceeding the social carrying capacity (tolerance) of ever-closer association with people because of property damage (such as reputed excessive accumulation of feces on walkways and athletic fields). Excessive geese numbers could impair human health and safety (via traffic hazards, the transmission of diseases, and attacks on people near the nest). Geese have also been impugned for damage to agricultural crops and contamination of water supplies.

However, the goose's tameness, conspicuousness, and dependence on humans for open and pseudo-tundralike habitats—such as areas of close-cropped grass—make it an ideal representative through which people of all ages can experience a close relationship with wildlife, and perhaps reestablish their psychological and physical roots in nature.

With respect to the latter, geese are also popularly hunted for food. In the Atlantic Flyway there are now about 1.1 million resident geese, and the Federal Canada Goose Management Plan has a goal of reducing that population to 650,000 birds by 2005. The U.S. Fish and Wildlife Service also has the ambitious goal of slashing the U.S. goose population. It is expected that this reduced population size will provide an optimal balance between geese's aesthetic value and the negative one associated with a too high population density.

According to the Atlantic Flyway Resident Canada Goose Management Plan (1999), it will not be possible to effectively reduce resident goose populations by egg treatment (shaking or oiling to make infertile) alone. Only a small percentage of the nests could be located, and even if 95 percent of all eggs found were treated each year, it is estimated that the result would still only be a 25 percent reduction in the goose population in ten years. Current hunting would have to increase about 50 percent, to a take of about 275,000 birds per year, to stabilize the current population. The report (Canada Goose Committee 1999, p. 20) notes: "It is unlikely that substantially higher harvests can be achieved with current hunter numbers and current regulations . . ." and suggests that "given the current status of resident geese, and the growing demand for relief of goose damage and conflicts—at a minimum, the need for federal permits for nest destruction, egg treatment, and shooting or capturing of small numbers of geese causing damage during spring or summer should be eliminated. This would reduce the administrative burden on state and federal agencies and property owners experiencing damage or conflict with resident geese."

An alternative to increased sports hunting is capturing geese during the summer flightless period in problem areas and processing the birds by local food bank programs, especially in urban areas. Such programs have already been successfully tried in New York, Virginia, Delaware, Maryland, Rhode Island, and Connecticut. The flightless birds group into flocks and it is easy to herd and capture them. Personally, I feel it would not be a bad thing if geese cover golf courses with a foot-deep layer of guano. That theoretical scenario, if it could be achieved, would be a small price to pay (for me!) for the grand spectacle of millions of geese flying untrammeled through the skies. But that may be a minority opinion.

If my little exercise in geese-watching has taught me one thing, it is that geese would eventually manage their own population as a result of their intense competition and their mutual destruction of eggs and young at breeding sites. Hunting pressure would, in general, have little effect on the population, because it would simply relax the competition and strife that would otherwise limit reproduction. This is not to say that a lot of hunting would not do it harm. It all depends on the situation, and I'm not qualified to speculate about that, nor do I want to claim the moral high ground and declare that hunting is wrong. But that may just be because it is easy for me—for days on end I can just sit and watch the action and enjoy it without firing a shot.

My bog contains one pair of geese, or at the most two. The geese themselves take care of the surplus. But they can't take care of the bog. The bog is a community of blue irises, sedges, marsh marigolds, and hundreds of other plant species. It is snipe, redwings, grackles, and dozens of other birds. It is beavers, mink, muskrats and water shrews, sunfish, minnows, and catfish. It is snapping and painted turtles, bullfrogs, spring peepers, and a dozen more amphibians. The bog's economy is run by bumblebees, white-bellied dragonflies, shiny green leaf beetles—hundreds of other insect species and thousands of small invisible players. The ducks and the geese are the visible tip. And it is hunters and former hunters who as a group have done the most to protect the swamps and other wetlands from being turned into cornfields.

Some say cornfields are a more efficient way to get food than eating meat. A standard rationalization holds that it's better to be a corn-eater than a meat-eater because you can get a lot of food out of a small corn patch but you need a huge area to raise the same amount of food in terms of meat.

Meat is indeed higher on the food chain. If you eat corn-fed beef you use up more corn than if you eat corn directly and you subsidize more cornfields that eat up swamps. Thus, the argument seems rational, but only so long as you eat beef or other corn-fed *domestic* animals (most are). But what if you eat *bison* from well-monitored and -regulated herds on the prairie? The answer is simple: You support prairie (that's also meadowlarks, etc.). Similarly, I do not begrudge hunters taking geese, especially those geese that are a result of population explosions due to man's interference. To get their wild geese, hunters will subsidize tundra, swamps, and other wetlands where they breed. We only protect what we need, and what we love.

A goose has, of course, a right to life (as does a hawk and a snipe, a weasel and a vole). But as David Ehrenfeld points out in his wonderful book *Beginning Again: People and Nature in the New Millennium*, rights conflict, and in that situation rights—ours included—imply obligation and they must defer to a higher authority. In the world of living organisms that higher authority is the ecosystem, the integrated complex of thousands of other species.

Aboriginal man and other predators have hunted geese and perhaps inadvertently maintained the ecological balance for aeons. We had little power or choice to do otherwise. The mass rearing of animals solely for slaughter is a recent phenomenon. But why keep animals raised for slaughter in pens when they could live freely? Ironic as it may seem, it may have something to do with salving our conscience. It seems morally repugnant to eat a majestic wild animal. It therefore makes sense to take that majestic animal out of nature, to pen it up as a prep for plastic wrap. Blindness may balm many a conscience, but it does not buy bread.

Wild geese provide many ways of connecting us to the land, to wild nature. The more ways we have and the more different kinds of interests and needs we can engage, the more we will try to maintain vigorous animal populations, and thus the beaver bogs that support them.

Appendix 4: Birds Seen at and around the Beaver Bog

B = breeding in the bog; * = single sighting

* Horned grebe, *Podiceps auritus*

* Pied-billed grebe, *Podilymbus podiceps*

American bittern, *Botaurus lentiginosus*

Great blue heron, *Ardea herodias*

B Canada goose, *Branta canadensis*

B Mallard, *Anas platyrhynchos*

American black duck, *Anas rubripes*

Green-winged teal, *Anas crecca*

Blue-winged teal, *Anas discos*

Wood duck, *Aix sponsa*

Ring-necked duck, *Aythya collaris*

Lesser scaup, *Aytha affinis*

Common goldeneye, *Bucephala clangula*

Bufflehead, *Bucephala albeola*

Common merganser, *Mergus merganser*

* Hooded merganser, *Lophodytes cucullatus*

B Virginia rail, *Rallus limicola*

Killdeer, *Charadrius vociferus*

* Solitary sandpiper, *Tringa solitaria*

Spotted sandpiper, *Actitis macularia*

B Common snipe, *Gallinago gallinago*

B American woodcock, *Scolopax minor*

* Bald eagle, *Haliaeetus leucocephalus*

Northern harrier, *Circus cyaneus*

Sharp-shinned hawk, *Accipiter striatus*

Broad-winged hawk, *Buteo platypterus*

Osprey, *Pandion haliaetus*

Ruffed grouse, *Bonasa umbellus*

Wild turkey, *Meleagris gallopavo*

B Mourning dove, *Zenaida macroura*

B Black-billed cuckoo, *Coccyzus erythropthalmus*

Barred owl, *Strix varia*

B Ruby-throated hummingbird, *Archilochus colubris*

Belted kingfisher, *Ceryle alcyon*

Northern flicker, *Colaptes auratus*

Yellow-bellied sapsucker, *Sphyrapicus varius*

Downy woodpecker, *Picoides pubescens*

Hairy woodpecker, *Picoides villosus*

Pileated woodpecker, *Dryocopus pileatus*

B Eastern kingbird, *Tyrannus tyrannus*

Olive-sided flycatcher, *Contopus borealis*

Eastern phoebe, *Sayornis phoebe*

B Least flycatcher, *Empidonax minimus*

B Willow flycatcher, *Empidonax traillii*

B Alder flycatcher, *Empidonax alnorum*

B Tree swallow, *Tachycineta bicolor*

Bank swallow, *Riparia riparia*

Barn swallow, *Hirundo rustica*

B Blue jay, *Cyanocitta cristata*

American crow, *Corvus brachyrhynchos*

Common raven, *Corvus corax*

Tufted titmouse, *Parus bicolor*

B Black-capped chickadee, *Parus atricapillus*

Winter wren, *Troglodytes troglodytes*

Golden-crowned kinglet, *Regulus satrapa*

Ruby-crowned kinglet, *Regulus calendula*

* Blue-gray gnatcatcher, *Polioptila caerulea*

B American robin, *Turdus migratorius*

B Veery, *Catharus fuscescens*

Northern shrike, *Lanius excubitor*

B Gray catbird, *Dumetella carolinensis*

Bohemian waxwing, *Bombycilla garrulous*

B Cedar waxwing, *Bombycilla cedrorum*

Red-eyed vireo, *Vireo olivaceus*

Warbling vireo, *Vireo gilvus*

Tennessee warbler, *Vermivora peregrina*

Nashville warbler, *Vermivora ruficapilla*

B Chestnut-sided warbler, *Dendroica pensylvanica*

Yellow-rumped warbler, *Dendroica coronata*

Palm warbler, *Dendroica palmarum*

B Yellow warbler, *Dendroica petechia*

B Northern waterthrush, *Seiurus noveboracensis*

B Common yellowthroat, *Geothlypis trichas*

B American redstart, *Setophaga ruticilla*

B Rose-breasted grosbeak, *Pheucticus ludovicianus*

B Northern cardinal, *Cardinalis cardinalis*

B Song sparrow, *Melospiza melodia*

American tree sparrow, *Spizella arborea*

Dark-eyed junco, *Junco hyemalis*

White-throated sparrow, *Zonotrichia albicollis*

White-crowned sparrow, *Zonotrichia leucophrys*

Fox sparrow, *Passerella iliaca*

B Swamp sparrow, *Melospiza georgiana*

B Red-winged blackbird, *Agelaius phoeniceus*

B Brown-headed cowbird, *Molothrus ater*

B Common grackle, *Quiscalus quiscula*

B Northern oriole, *Icterus galbula*

B American goldfinch, *Carduelis tristis*

Pine grosbeak, *Pinicola enucleator*

Common redpoll, *Carduelis flammea*

Purple finch, *Carpodacus purpureus*

Evening grosbeak, *Coccothraustes vespertinus*

Selected References

A. WETLANDS AND BEAVERS

Johnson, Charles W. 1985. *Bogs of the Northeast.* Hanover, N.H.: University of New England Press.

McDowell, L. L., R. M. Dole Jr., M. Howard Jr., and R. A. Farrington. 1971. Palynology and radiocarbon chronology of Bugbee Wildflower Sanctuary and Natural Area, Caledonia County, Vermont.

Müller-Schwarze, D., and L. Sun. 2003. *The Beaver: Natural History of a Wetland Engineer.* Ithaca, N.Y., and London: Cornell University Press.

Rue, Leonard Lee III. 1964. *The World of the Beaver.* Philadelphia and New York: J. B. Lippincott Co.

Seton, Ernest Thompson. 1953. *Lives of Game Animals.* Boston: Charles T. Bradford Co.

Whitehead, D., and D. R. Bentley. 1963. A post-glacial pollen diagram from southwestern Vermont. *Pollen et Spores* V:115–27.

B. LORENZ'S GEESE

Deichmann, Ute (translated by Thomas Dunlap). 1996. *Biologists under Hitler.* Cambridge, Mass.: Harvard University Press.

Heinroth, O. 1910. Beiträge zur Biologie, namentlich Ethologie und Psychologie der Anatiden. *Verh. 5 Int. Ornith. Kongr. Berlin* 589–702.

Hess, E. 1972. "Imprinting" in a natural laboratory. *Sci. Am.* 277(8): 24–31.

Kalikow, T. J. 1980. *Die ethologische Theorien von Konrad Lorenz: Erklaerung und Ideologie, 1938 bis 1943*. In *Naturwissenschaften und Technie im Dritten Reich*, edited by H. Mehrtense and S. Richter. Frankfurt: Suhrkamp.

———. Konrad Lorenz's ethological theory: Explanation and ideology, 1938–1943. *J. Hist. Biol.* 16:39-72.

Klopfer, Peter, H. 1999. *Politics and People in Ethology*. Associated University Presses. Cranbury, N.J.

Lorenz, K. Z. *The Foundations of Ethology*. New York: Springer-Verlag.

Lorenz, K. 1940. Durch Domestikation verursachte Stoerungen arteigenen Verhaltens. *Z. Angew. Psychol.* 59:2-82.

———. 1988. *Hier bin ich—wo bist du? Ethologie der Graugans.* Munich: Piper.

Tinbergen, N., and K. Lorenz. 1938. Taxis und Instinkthandlung in der Eirollbewegung der Graugans. *Z. Tierpsychol.* 1/2: 1–29.

C. CANADA GEESE

Abraham, K. F., J. O. Leafloor, and D. H. Rusch. 1999. Molt migrant Canada geese in northern Ontario and western James Bay. *J. Wildl. Mgt.* 63:649–55.

Ankney, C. D. 1996. An embarrassment of riches: too many geese. *J. Wildl. Mgt.* 60:217–23.

Bellrose, F. C. 1976. *Ducks, Geese and Swans of North America.* Harrisburg, Pa.: Stackpole Books.

Benson, D., S. Browne, and J. Moser. 1982. Evaluation of hand-reared goose stocking. Final Rep. Fed. Aid Project W-39-R, Job No. IV-2, N.Y. State Dept. Environ. Conserv., Bureau of Wildlife, Delmar. 24 pp.

Bent, A. C. 1925. Life histories of North American wild fowl. *W.S. Nat. Mus. Bull.* 130:316.

Bertness, M., B. R. Silliman, and R. Jeffries. 2004. Salt marshes under siege. *American Scientist.* 92: 54–61.

Canada Goose Committee. 1999. Atlantic Flyway Resident Canada Goose Management Plan. 42 pp.

Collias, N. E., and L. Jahn. 1959. Social behavior and breeding success in Canada geese *(Branta canadensis)* confined under seminatural conditions. *Auk* 76:478–509.

Conover, M. R., and G. G. Chasko. 1985. Nuisance Canada goose problems in the eastern United States. *Wildl. Soc. Bull.* 13:228–33.

Converse, K. A., and J. J Kennelly. 1994. Evaluation of Canada goose sterilization for population control. *Wildl. Soc. Bull.* 22:265–69.

Craighead, F. C., and J. J. Craighead. 1949. Nesting Canada geese on the Upper Snake River. *J. Wildl. Mgt.* 13:51–64.

Craighead, J. J., and D. S. Stockstad. 1964. Breeding age of Canada geese. *J. Wildl. Mgt.* 28:57–64.

Dill, H. H., and F. B. Lee, eds. 1970. *Home Grown Honkers.* U.S. Fish and Wildl. Serv., Washington, D.C. 154 pp.

Gosser, A. L., and M. R. Conover. 1999. Will the availability of insular nesting sites limit reproduction in urban Canada goose populations? *J. Wildl. Mgt.* 63:369–73.

Hanson, H. C. 1949. Notes on white spotting and other plumage variations in geese. *Auk* 66:164–71.

———. 1953. Interfamily dominance in Canada geese *(Branta canadensis interior). Auk* 70:11–16.

———. 1959. The incubation patch of wild geese: its recognition and significance. *Arctic* 12:139–50.

———. 1997. *The Giant Canada Goose.* Carbondale and Edwardsville: Southern Illinois University Press.

Hanson, H. C., and C. Currie. 1957. The kill of wild geese by the natives of the Hudson-James Bay region. *Arctic* 10:211–29.

Hanson, H. C., and R. L. Browning. 1959. Nesting studies of Canada geese on the Hanford Reservation, 1953–1956. *J. Wildl. Mgt.* 23:129–37.

Hestbeck, J. B., J. D. Nichols, and R. A. Malecki. 1991. Estimates of movement and site fidelity using mark-resight data of wintering Canada geese. *Ecology* 72:523–33.

Jamieson, R. L. 1998. Tests show Canada geese are cause of polluted lake water. *Seattle Pilot* July 9.

Jenkins, D. W. 1944. Territory as a result of despotism and social organization in geese. *Auk* 61:30–47.

Johnson, F. A., and P. M. Castelli. 1998. Demographics of "resident" Canada geese in the Atlantic Flyway. Pages 127–33 in D. H. Rusch, M. D. Samuel, D. D. Humburg, and B. D. Sullivan, eds. Biology and management of Canada geese. *Proc. Int. Canada Goose Symp.*, Milwaukee, Wis.

Keefe, T. 1996. Feasibility study on processing nuisance Canada geese for human consumption. Minn. Dept. Natur. Resour., Forest Lake, Minn. 7 pp.

Kossack, C. W. 1947. Incubation temperatures of Canada geese. *J. Wildl. Mgt.* 11:119–26.

Kuyt, E. 1962. Northward dispersion of banded Canada geese. *Can. Field-Naturalist* 76:180–81.

Nelson, H. K., and R. B. Oetting. 1998. Giant Canada goose flocks in the United States. Pages 127–33 in D. H. Rusch, M. D. Samuel, D. D. Humburg, and B. D. Sullivan, eds. Biology and management of Canada geese. *Proc. Int. Canada Goose Symp.*, Milwaukee, Wis.

Pearson, T. G., ed. 1917. *Birds of America.* Garden City, N.Y.: Garden City Books.

Pottie, J. J., and H. W. Heusmann. 1979. Taxonomy of resident Canada geese in Massachusetts. *Trans. Northeast Fish and Wildl. Conf.* 36:132–37.

Serie, J., and A. Vecchio. 1999. Atlantic Flyway Midwinter Waterfowl Survey, 1999, Final Report. U.S. Fish and Wildl. Serv., Laurel, Md. 2 pp.

Sheaffer, S. E., and R. A. Malecki. 1998. Status of Atlantic Flyway resident nesting Canada geese. Pages 29–34 in D. H. Rusch, M. D. Samuel, D. D. Humburg, and B. D. Sullivan, eds. Biology and management of Canada geese. *Proc. Int. Canada Goose Symp.*, Milwaukee, Wis.

Sheaffer, S. E., R. A. Malecki, and R. E. Trost. 1987. Survival, harvest, and distribution of resident Canada geese in New York, 1975–84. *Trans. NE Sect. Wildl. Soc.* 44:53–60.

Williams, C. S., and M. C. Nelson. 1943. Canada goose nests and eggs. *Auk* 60:341–45.

Wood, J. S. 1964. Normal development and causes of reproductive failure in Canada geese. *J. Wildl. Mgt.* 28:197–208.

Yocum, C. F. 1952. Techniques used to increase nesting of Canada geese. *J. Wildl. Mgt.* 16:425–28.

D. BROOD PARASITISM AND NEST PREDATION

Andersson, M. 1984. Brood parasitism within species. In *Producers and Scroungers: Strategies of Exploitation and Parasitism,* edited by C. J. Barnard. London: Croom Helm. Pp. 195–228.

Anderson, V. R., and R. T. Alisauskas. 2002. Composition and growth of King Eider ducklings in relation to egg size. *Auk* 119:62–70.

Arnold, T. W., D. W. Howertar, J. H. Devries, B. L. Joynt, R. B. Emery, and M. G. Anderson. 2002. Continuous laying and clutch size in mallards. *Auk* 119:261–66.

Eadie, J. M., F. P. Kehoe, and T. D. Nudds. 1988. Pre-hatch and post-hatch brood amalgamation [in *North American Anatidae*]: a review of hypotheses. *Can. J. Zool.* 66:1709–21.

Payne, R. B. 1977. The ecology of brood parasitism in birds. *Ann. Rev. Ecol. & Syst.* 8:1–28.

Sherman, P. W. 2001. Wood ducks: a model system for investigating conspecific parasitism in cavity-nesting birds. In *Model Systems in Behavioral Ecology,* edited by L. A. Dugatkin. Princeton, N. J.: Princeton University Press. Pp. 311–37.

Sorenson, M. D. 1991. The functional significance of parasitic egg laying and typical nesting in redhead ducks: an analysis of individual behavior. *Anim. Behav.* 42:771–96.

———. 1993. Parasitic egg laying in canvasbacks in frequency, success, and individual behavior. *Auk* 110:57–69.

———. 1998. Patterns of parasitic egg laying and typical nesting in redhead and canvasback ducks. In *Parasitic Birds and Their Host: Studies in Coevolution,* edited by S. I. Rothstein and S. K. Robinson. New York: Oxford University Press. Pp. 357–75.

Weigmann, C., and J. Lemprecht. 1991. Intra-specific nest parasitism in bar-headed geese, *Anser indicus. Anim. Behav.* 41:677–88.

Yom-Tov, Y. 1980. Intraspecific nest parasitism in birds. *Biol. Rev.* 55:93–108.

E. BLACKBIRDS AND COLONIALITY

Allen, A. A. 1914. The red-winged black bird: a study in the ecology of a cat-tail marsh. *Proc. Linn. Soc. N.Y.* 24–25:43–128.

Allen, C. S. 1892. Breeding habits of the fishhawk on Plum Island, New York. *Auk* 9:313–21.

Bent, A. C. 1965. *Life Histories of North American Blackbirds, Orioles, Tangers, and Allies.* New York, Dover Publications.

Foster, F. B. 1927. Grackles killing young pheasants. *Auk* 44:106.

Mayr, E. 1941. Red-wing observations of 1940. *Proc. Linn. Soc. N.Y.* 52–53:75–83.

Nero, R. W. 1984. *Redwings.* Washington, D.C.: Smithsonian Press.

Orians, G. H. 1980. *Some Adaptations of Marsh-Nesting Blackbirds.* Princeton, N.J.: Princeton University Press.

Picman, J. 1984. Experimental study on the role of intra- and interspecific competition in the evolution of nest-destroying behavior in marsh wrens. *Can. J. Zool.* 62:2353–56.

Picman, J., and A. Isabelle. 1995. Sources of nesting mortality and correlates of nesting success in yellow-headed blackbirds. *Auk* 112:183–91.

Picman, J., M. L. Leonard, and A. Horn. 1988. Antipredation role of clumped nesting by marsh-nesting red-winged blackbirds. *Behav. Ecol. & Sociobiol.* 22:9–15.

Picman, J., S. Pribil, and A. Isabelle. 2002. Antipredator value of colonial nesting in yellow-headed blackbirds. *Auk* 119(2):461–72.

Pribil, S., and J. Picman. 1994. Black-capped chickadees, *Parus atricapillus,* eat eggs of other birds. *Can. Field-Naturalist* 108:371–72.

Searcy, W. A., and K. Yasukawa. 1995. *Polygyny and Sexual Selection in Red-winged Blackbirds.* Princeton, N.J.: Princeton University Press.

Shipley, F. S. 1979. Predation on red-winged blackbird eggs and nestlings. *Wilson Bull.* 91:426–33.

Snelling, J. C. 1968. Overlap in feeding habits of redwing blackbirds and common grackles nesting in a cattail marsh. *Auk* 85:560–85.

Spooner, A., S. Pribil, and J. Picman. 1996. Why do gray catbirds destroy eggs in nests of other birds? *Can. J. Zool.* 74:1688–95.

Ward, P., and A. Zahavi. 1973. The importance of certain assemblages of birds as "information centers" for food finding. *Ibis* 115:517–34.

Weber, J. A. 1912. A case of cannibalism among blackbirds. *Auk* 29:394–95.

Wiens, J. A. 1965. Behavioral interactions of red-winged blackbirds and common grackles on a common breeding ground. *Auk* 82:356–74.

F. GENERAL

Conover, A. 1998. *Fly Away Home.* Smithsonian. April. Pp. 63–72.

Dupre, K. 2001. *The Raven's Gift: A True Story from Greenland.* Boston: Houghton Mifflin.

Ehrenfeld, D. 1993. *Beginning Again: People and Nature in the New Millennium.* New York and Oxford: Oxford University Press.

Hay, J. 1998. *In the Company of Light.* Boston: Beacon Press.

Kimber, R. 2002. *Living Wild: The Education of a Hunter-Gardener.* Guilford, Conn.: Lyons Press.

Lishman, W. 1996. *Father Goose.* New York: Crown Publishers.

Lorenz, K. 1978. *The Year of the Greylag Goose.* New York: Harcourt Brace Jovanovich.

Nice, M. M. 1964. *Studies in the Life History of the Song Sparrow.* Vol II. New York: Dover Publications. (First published in 1943 as Vol. IV of the *Transactions of the Linnaean Society of New York.*)

Saunders, A. A. 1924. Recognizing individual birds by song. *Auk* 41:242–49.

Tinbergen, N. 1989. *The Study of Instinct.* Oxford, U.K.: Oxford University Press. (First published in 1951.)

Wheatherhead, P. J. 1979. Ecological correlations of monogamy in tundra-breeding Savannah sparrows. *Auk* 96:391–401.

White, E. B. 1977. *Essays of E. B. White.* New York: Harper and Row.

Index

BOOKS BY BERND HEINRICH:

THE GEESE OF BEAVER BOG
ISBN 0-06-095738-7 (paperback)
With a biologist's eyes and a curious nature-lover's soul, Bernd Heinrich set out to the beaver bog adjacent to his house to observe and understand the daily life of Canadian geese—whose routines are as colorful and dramatic as those of their human counterparts.
"Arguably today's finest naturalist author. . . . This is another worthy missive from our latter-day Thoreau." —*Publishers Weekly*

WINTER WORLD: *The Ingenuity of Animal Survival*
ISBN 0-06-019744-7 (hardcover)
Examining everything from food sources in the barren landscape to the chemical composition that allows certain creatures to survive, Heinrich probes the mysteries by which nature sustains herself through the cruel exigencies of winter.
"Heinrich, who combines his keen scientific eye with the soul of a poet, enthralls with this new, captivating, and at times surprising examination of animal survival in the coldest of seasons." —*New York Times Book Review*

MIND OF THE RAVEN: *Investigations and Adventures with Wolf-Birds*
ISBN 0-06-093063-2 (paperback)
Bernd Heinrich finds himself dreaming of ravens and decides he must get to the truth about this animal reputed to be so intelligent.
"An amazing book. . . . Heinrich has documented a level of intelligence and social sophistication rarely even dreamed to exist in birds." —Edward O. Wilson

THE TREES IN MY FOREST
ISBN 0-06-092942-1 (paperback)
Biologist and acclaimed nature writer Bernd Heinrich takes readers on an eye-opening journey through the hidden life of a forest.
"In Heinrich's hands, the lives of trees are as noble and dramatic as the lives of men." —*Washington Post*

WHY WE RUN: *A Natural History*
ISBN 0-06-095870-7 (paperback)
ORIGINALLY PUBLISHED IN HARDCOVER AS *RACING THE ANTELOPE*
Weaving evolution, intelligence, and imagination with his own stories of long-distance running, Bernd Heinrich has combined the rigors of science with the passion of running.
"A remarkable perspective on how and why running is an integral part of what makes us human." —*UltraRunning* magazine